MICROSCOPY HANDBOOKS 40

Fluorescence Microscopy

SECOND EDITION

T0138971

Royal Microscopical Society MICROSCOPY HANDBOOKS

Fluorescence Microscopy

SECOND EDITION

B. Herman
University of North Carolina
Chapel Hill, NC 27599, USA

Taylor & Francis
Taylor & Francis Group

In association with the Royal Microscopical Society

Published by:

Taylor & Francis Group

In US: 270 Madison Avenue,
       New York, NY 10016

In UK: 2 Park Square, Milton Park
       Abingdon, Oxon OX14 4RN

First edition © 1987 Royal Microscopical Society
Second edition © 1998 Taylor & Francis Group

Transferred to digital printing 2006

First published 1987 (ISBN 0 19 856408 2)
Second edition 1998 (ISBN 1 872748 84 8)

This book contains information obtained from authentic and highly regarded sources.
Reprinted material is quoted with permission, and sources are indicated. A wide variety
of references are listed. Reasonable efforts have been made to publish reliable data and
information, but the author and the publisher cannot assume responsibility for the
validity of all materials or for the consequences of their use.

All rights reserved. No part of this book may be reprinted, reproduced, transmitted, or
utilized in any form by any electronic, mechanical, or other means, now known or
hereafter invented, including photocopying, microflming, and recording, or in any
information storage or retrieval system, without written permission from the
publishers.

A catalog record for this book is available from the British Library.

Production Editor: John Leonard
Typeset by Poole Typesetting (Wessex) Ltd, Bournemouth, UK.

10 9 8 7 6 5 4 3 2 1

Front cover: Alterations in cytosolic calcium in human epidermal keratinocytes following exposure
to epidermal growth factor (EGF). Keratinocytes were loaded with Fura-2 and exposed to 10 ng ml⁻¹
EGF. Warm colours (yellow, red) represent areas of high calcium, while cool colours (violet, blue)
represent areas of low calcium.

Publisher's Note
The publisher has gone to great lengths to ensure the quality of this reprint, but points out that
some imperfections in the original may be apparent.

Taylor & Francis Group, an informa business   Visit our web site at http://www.garlandscience.com

Contents

Abbreviations

A	absorption
AFIC	automated fluorescence image cytometry
A/D	analogue-to-digital converter
Ala	alanine
ALU	arithmetic logic unit
AM	acetoxymethyl
AMCA	amino-methyl-coumarin-acetic acid
AOD	acoustic-optical device
AOTF	acousto-optical tunable filter
ASA	film speed
BAPTA	O,O'-bis(2-aminophenyl)ethyleneglycol-N, N, N', N'-tetraacetic acid
BCECF	3'-O-acetyl-2',7'-bis(carboxyethyl)-4- or -5-carboxy-fluorescein
BCPDA	4,7-bis(chlorosaltophenol)-1,10-phenanthidine-2,9-dicarboxylic acid
BFP	blue fluorescent protein
c	concentration
CAT	computerized axial tomography
CBS or DBS	dichromatic beam splitter
CCD	charge coupled device
CD ROM	compact disk read only material
cDNA	complementary DNA
CID	change injector device
CMYK	cyan, magenta, yellow, black
CPU	central processing unit
CRT	cathode ray tube
CW	continuous wave
CY3	cyanine 3
D	diffusion coefficient
DAPI	4',6-diamidino-2-phenylindole
DIC	differential interference contrast
DNA	deoxyribonucleic acid
dNTP	deoxynucleotide trisphosphate
DTNB	5,5'-Dithiobis(2-nitrobenzoic acid)
DTT	dithiothreitol

DVM	digitized video microscopy
EGF	epidermal growth factor
EGFP	enhanced GFP
EGTA	O,O'-Bis(2-aminoethyl)ethyleneglycol-N, N, N', N'-tetraacetic acid
EM	electron microscopy
EPR	exhaustive photon reassignment
E_T	efficiency of FRET
F/P	fluorophore/protein molar ratio
FAD	flavin adenine dinucleotide
fc	foot candles
FEP	fluorescence emission before photodestruction
FER	fluorescence emission rate
FISH	fluorescence *in situ* hybridization
FITC	fluorescein isothiocyanate
FL	fluorophore
FLIM	fluorescence lifetime imaging microscopy
FMN	flavin mononucleotide
FRAP	fluorescence recovery after photobleaching
FRET	fluorescence resonance energy transfer
FRETM	fluorescence resonance energy transfer microscopy
FWHM	full width half maximum
GFP	green fluorescent protein
Gly	glycine
G_o	ground state
HPV	human papillomavirus
HSI	hue, saturation, intensity
I_A	acceptor emission
I_D	donor emission
I_O	incident intensity
I	observed intensity
ISIT	intensified silicon intensified target
ISO	Industrial Standards Organization
$J(\lambda)$	overlap integral
K^2	orientation factor
k_f	rate constant for fluorescence emission
k_i	rate constant for internal conversion
K_M	Michaelis–Menton constant
k_{nr}	rate constant for non-radiative processes
k_p	rate constant for phosphorescence
K_T	rate constant for RET
LCTF	liquid crystal tunable filter
LM	light microscopy
LP	long pass
LSCM	laser scanning confocal microscope
LSM	laser scanning microscope
Mag	magnification

MCP	multichannel plate
MDVM	multiparameter digitized video microscopy
Met	methionine
mf	mobile fraction
MOS	metal oxide semiconductor
mt	mutant
NA	numerical aperture
ND	neutral density
NIH	National Institute of Health
NITR-5	caged calcium
NITR-7	caged calcium
OMDR	optical memory disk recorder
OPD	optical path difference
p	measured polarization
PBS	phosphate-buffered saline
PCB	printed circuit board
PCR	polymerase chain reaction
Phe	phenylalanine
PMT	photomultiplier tube
p_0	limiting polarization
Q	quantum yield
r	measured anisotropy
r_{Airy}	radius of Airy disc
r_0	limiting anisotropy
RET	resonance energy transfer
RGB	red, green, blue
S/N	signal-to-noise ratio
S_1	singlet excited state
SDS	sodium dodecyl sulphate
Ser	serine
SIT	silicon intensified target
SLR	single lens reflex
S_0	ground state
SP	short pass
SSCM	stage scanning confocal microscope
TE	transverse electronic
TeO_2	tellurium oxide
Thr	threonine
TIFF	tagged image file format
TIRF	total internal reflectance fluorescence microscopy
T_m	melting temperature
TM	transverse magnetic
TPEM	two-photon excitation microscopy
Tyr	tyrosine
Val	valine
wt	wild-type
x	pathlength

ε	extinction coefficient
η	refractive index of medium
λ	wavelength
μlux	microlux
ν_1	vibrational energy level of excited state
ν_0	vibrational energy level of ground state
φ	rotational correlation time
τ	fluorescence lifetime
τ_D	lifetime of donor
τ_{DA}	lifetime of donor in presence of acceptor
τ_F	fluorescence lifetime
τ_0	intrinsic fluorescent lifetime

Preface

Fluorescence microscopy is a very powerful tool in that it allows quantitative spatial and temporal visualization of fluorescent material in microscope specimens. Because of this capability, numerous investigators have turned to this technology to address questions of fundamental biological importance. Having used fluorescence microscopy in my own research, I am often asked an array of questions about this technology, and especially, where individuals can turn to learn more about this subject. While many excellent texts and monographs exist which cover various aspects of fluorescence and microscopy, the lack of a concise comprehensive source for this information was a major motivation for writing this book. I hope to have relayed my understanding of the important principles and applications of fluorescence microscopy, and that this book will serve as a broad introduction for new users to the variety of information that can be gathered with fluorescence microscopy.

Much of the information presented in this book is the result of the substantial contributions of many of my colleagues. While the list is long, particularly important contributions were made by David Albertini, Richard Berlin, John Bogan, Ludwig Brand, Robert Clegg, Jim DiGuiseppi, Pamela Diliberto, Fred Fay, Salvatore Fernandez, Kathryn Florine-Casteel, Hans Gerritsen, Gerry Gordon, Enrico Gratton, Charles R. Hackenbrock, Richard Haugland, Richard Inman, Shinya Inoué, Colin Izzard, Ken Jacobson, Tom and Donna Jovin, Joe Kao, Hans Kapitza, Ernst Keller, Dennis Koppel, Joseph Lakowicz, John Lederer, John J. Lemasters, Leslie Loew, Steve Lockett, Fred Maxfield, Butch Moomaw, Stephen Morris, John Murray, Ammasi Periasamy, David Piston, Ted Salmon, Jan Slavik, Deborah K. Smith, Ken Spring, Hans Tanke, D. Lansing Taylor, Roger Tsien, Jim Turner, Xue Feng Wang, Yu-Li Wang, Pawel Wodnicki and Barney Wray. The support of numerous funding agencies and corporations is also appreciated, including the NIH, NSF, ACS, AHA, The North Carolina Biotechnology Center, and the Gustavus and Louise Pfeiffer Foundation. Dage-MTI Inc., Datacube, Inc., Hamamatsu, Inc., Olympus, Inc. and Carl Zeiss, Inc., are also gratefully acknowledged. This book would not be a reality without the work of an exceptionally talented graphic artist, Tracey Curran. Lastly, I wish to dedicate this book to my wife, Deborah K. Smith, and our beautiful daughter, Lindsey, without whose understanding, love and patience I would not have been able to undertake this endeavour.

Brian Herman

1 Fundamentals of fluorescence

1.1 What is fluorescence?

Fluorescence is the property of some atoms and molecules to absorb light of a particular wavelength and after a brief interval, termed the fluorescence lifetime, to re-emit light at longer wavelengths. Fluorescence requires an outside source of energy, is the result of the absorption of light, and involves the emission of electromagnetic radiation (light). This process is different from chemiluminescence, where the excited state is created via a chemical reaction.

1.2 Wavelength, excitation and emission spectra

Light can be described as having characteristics of both particle and wave phenomena (*Figure 1.1*).

As a wave, light has two components, electric (E) and magnetic (H), which travel in space perpendicular to one another. The distance between consecutive wavecrests is defined as the wavelength (λ). Light normally consists of a mixture of electromagnetic waves of many wavelengths. Planck's law states that $E = h\upsilon$, where E is the energy in ergs, h is Planck's constant (6.6×10^{-27}) and υ is the frequency of light (sec^{-1}; i.e. the number of waves passing a point in 1 sec). As a reference, the energy of 1 mole of photons (6.02×10^{23}) at 500 nm is ~ 60 kcal mol^{-1}. Shorter wavelengths (i.e. shorter distances between consecutive wavecrests) have higher amounts of energy versus longer wavelengths.

The absorption of a photon of energy by a fluorescent molecule is an all or none phenomenon and each fluorescent molecule can only absorb incident light of certain specific wavelengths known as absorption

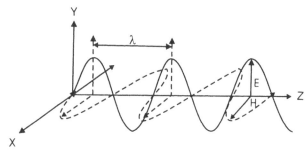

Figure 1.1. An electromagnetic light wave with electric field vector **E** in the *yz* plane and magnetic field vector **H** in the *xz* plane. The distance between consecutive wavecrests is defined as the wavelength, λ. Reproduced by permission of Marcel Dekker, New York, from Pesce, A.J., Rosen, C.-G. and Pashby, T. (1971) *Fluorescence Spectroscopy: An Introduction for Biology and Medicine*, p. 3.

bands. Emission of fluorescence also occurs at certain specific wavelengths, but these wavelengths are at lower energies (i.e. longer λ values) than absorption band maxima due to loss of energy by the molecule during interactions with its environment before it emits its fluorescence (internal conversion).

Absorption of energy by fluorescent molecules occurs between a number of closely spaced vibrational and rotational excited states in different orbitals. The Jablonski diagram (named after A. Jablonski and described in 1953) demonstrates the different energy levels involved in the absorption and emission of light (*Figure 1.2*). Physically, absorption of light occurs very quickly (approx. 10^{-15} sec) and corresponds to the excitation of the fluorophore from the ground state to an excited state. Relaxation to the lowest excited singlet state, known as internal conversion, occurs within approximately 10^{-11} sec as energy is thermally transferred to the environment. Internal conversion (loss of energy in the absence of emission of light) is due to collision of the excited state probe with solvent molecules. The molecule 'lives' in the lowest excited singlet state for periods of the order of nanoseconds (approx. 10^{-9} sec). Relaxation from this state (the lowest excited singlet, S_1; see *Figure 1.2*) to the ground state with emission of a photon is, physically, what is referred to as fluorescence. Each fluorescent molecule (fluorophore) can repeat the excitation/emission process many times, for example for fluorescein isothiocyanate (FITC) approximately 30 000 times, before excited state processes bleach the FITC molecule. In addition to fluorescence, molecules which reside in the lowest excited singlet state can undergo intersystem crossing to the triplet state from which a longed lived emission, phosphorescence, occurs. Fortunately, this is a relatively rare event. Delayed fluorescence can also occur due to transitions from T_1 back to S_1 and then to G_0.

The probability of movement of an electron from the ground state (G_0) to the excited state (S_1) depends on the degree of similarity of the vibrational and rotational energy states where the electron resides in the ground state versus where it would reside in the excited state (*Figure*

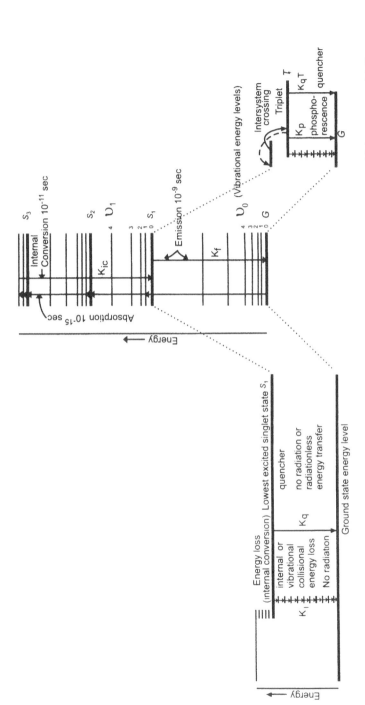

Figure 1.2. Modified Jablonski diagram illustrating the lifetimes of electronic transitions and the various processes which compete with fluorescence (k_f) (k_i = internal conversion, k_q = quenching, k_{pT} = intersystem crossing into triplet state, k_p = phosphorescence) for deactivation of excited energy state. υ_0 and υ_1 represent vibrational energy states (*see Figure 1.3*).

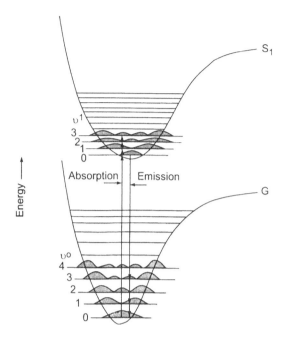

Figure 1.3. Energy diagram. Shaded areas represent the probability of the electron position in the ground (υ_0) and excited (υ_1) vibrational energy states. Transitions from the ground to the excited state occur in such a short time (10^{-15} sec) that the molecule cannot move (Frank Condon principle). Therefore, the only transitions from the ground state to the excited state which can occur are those where the probability of the electron position in the ground and excited states maximally overlap. Reproduced by permission of Marcel Dekker, New York, from Pesce, A.J., Rosen, C.-J. and Pashby, T. (1971) *Fluorescence Spectroscopy: An Introduction for Biology and Medicine*, p. 42.

1.3). The most favoured electron transitions will be those where the probability of the position of the electron in the ground and excited states (rotational and vibrational) maximally overlap. This position of maximal overlap can be thought of as indicating the excitation energy level that is most likely to be absorbed. The most likely state for an electron at room temperature is the ground state (G_0 or S_0). Within the ground state there exist a number of distinct vibrational energy states ($\upsilon_0 = 0, 1, 2$), and thus the amount of energy of absorption for each of these vibrational levels in the ground state will differ. This gives rise to the absorption spectrum containing multiple peaks (*Figure 1.4*). If we scan through the absorption spectrum while looking at the emission at one wavelength, we generate the excitation spectrum (*Figure 1.5*). Like the absorption spectrum, the excitation spectrum is broadened. If we hold the excitation constant and scan through the emission wavelengths, we generate the emission spectrum (*Figure 1.5*).

Following the absorption of energy and movement of the electron to the excited state, internal conversion causes the excited state electron to lose energy, such that the electron comes to rest in the lowest vibrational energy level of the excited singlet state ($\upsilon_1 = 0$; *Figure 1.3*). The energy

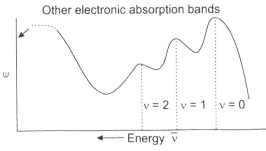

Other electronic absorption bands

ε

v = 2   v = 1   v = 0

Energy ⊽

Figure 1.4. Electronic absorption bands of a hypothetical fluorophore where there is little separation between the vibrational energy levels of the ground state (see *Figure 1.3*). Adapted with permission from Marcel Dekker, New York, from Pesce, A.J., Rosen, C.-J. and Pashby, T. (1971) *Fluorescence Spectroscopy: An Introduction for Biology and Medicine,* p. 43.

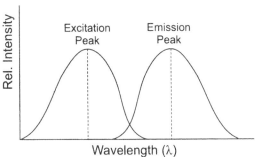

Rel. Intensity

Excitation Peak

Emission Peak

Wavelength (λ)

Figure 1.5. Typical excitation and emission spectrum of a fluorophore.

emitted by the photon then will be the difference between the energy level of the $v_1 = 0$ state and that to whichever ground state the electron falls.

1.2.1 Mirror image rule

As with absorption, the probability that an excited state electron will return to a particular ground state vibrational energy levels is proportional to the overlap of the electron vibrational and rotational energy levels of the ground and excited state energy levels (*Figure 1.3*). In the absence of other competing processes, this effectively results in the fluorescence emission spectrum being a mirror image of the absorption spectrum; that is, the probability of returning to a particular energy level in the ground state is similar to the position the molecule occupied in the ground state before excitation (*Figure 1.6*). This is known as the mirror image rule.

1.2.2 Stokes' shift

'Stokes' shift' refers to the fact that emission is generally shifted to longer wavelengths than the absorption because emission occurs from lower energy singlet states (*Figures 1.6* and *1.7*). The existence of the Stokes' shift is critical to making fluorescence such a sensitive technique. Stokes' shift allows a low number of emitted photons to be detected against a low background separated from the large number of excitation photons.

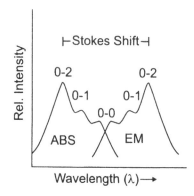

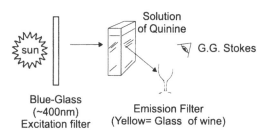

Figure 1.6. Mirror image rule. The probability of the electron returning to a particular ground state vibrational energy level is related to the probability of the electron position in the ground state before excitation (i.e. the most likely transition would be as follows: $\upsilon_0 = 0 \to \upsilon_1 = 2 \to \upsilon_1 = 0 \to \upsilon_0 = 2$). This results in the emission spectrum being a mirror image of the absorption spectrum.

Figure 1.7. Observation of Stokes' shift. Adapted with permission of Plenum Press, New York from Lakowicz, J.R. (1993) *Principles of Fluorescence Spectroscopy*.

1.3 Quantum yield

The quantum yield, Q, is a measure of the emission efficiency of the fluorophore and is expressed as the ratio of photons emitted to photons absorbed:

$$Q = \frac{\text{photons emitted}}{\text{photons absorbed}}$$

Q can range from 0 to 1.

Fluorescence is proportional to light absorbed multiplied by the quantum yield of the fluorophore. The following relation gives us some insight:

$$F = I_0 \varepsilon [c] x Q,$$

where I_0 is the intensity of the light beam falling on the solution, ε the molar extinction of the chromophore, $[c]$ the concentration of the chromophore and x the pathlength of the beam through the solution. The value ε is usually expressed at the maximum wavelength of absorption, while Q is a measure of the total photon emission over the entire fluorescence spectrum. Thus, the amount of fluorescence is proportional to the product of the light absorbed ($I_0 \varepsilon [c] x$) by the fluorophore and the quantum yield, Q. The term $I_0 \varepsilon [c] x$ is a result of Beer's law for low optical densities, namely that the amount of light absorbed by a set of fluorophores is proportional to the incident light intensity [$A = -\log (I/I_0)$], extinction coefficient, quantum yield and pathlength of the fluorophore solution. From this relationship, if I_0, ε, x and Q are known it is possible to determine $[c]$ as a function of F.

1.4   Fluorescence lifetime

The fluorescence lifetime, τ, is defined as the characteristic time that a molecule remains in an excited state prior to returning to the ground state. During the lifetime of the excited state, the fluorophore can undergo conformational changes as well as interacting with its local environment. If a uniform population of fluorescent molecules is excited with a brief pulse of excitation light, the decay of fluorescence intensity as a function of time is described by the exponential function:

$$I(t) = I_0 \, e^{(-t/\tau)}$$

where $I(t)$ is the intensity measured at time t, I_0 is the initial intensity immediately after the excitation pulse and τ is the fluorescence lifetime. The fluorescence lifetime is defined as the time in which the fluorescence intensity decays to 1/e of the initial intensity (i.e. 37% of I_0) (*Figure 1.8*). The fluorescence lifetime is the inverse of the rate constant of the decay of fluorescence from the excited state to the ground state ($1/k_f$; see *Figure 1.2*).

Since the amount of fluorescence is proportional to the excited state population of singlets, to measure the fluorescent lifetime, we can interrogate the population by asking how fast fluorescence decays after a pulse of excitation (*Figure 1.8*). The decay of a fluorescent molecule in a uniform solvent is usually monoexponential. In cells, where multiple environments exist, the decay of the fluorescent lifetime of a fluorophore is often multiexponential. A number of other deactivation or energy depleting processes can compete with fluorescence for return of the excited state electrons to the ground state. These include (*Figure 1.2*): internal conversion (k_i), phosphorescence (k_p) and quenching (k_q). Other than fluorescence and phosphorescence, the processes for return of the excited state electrons to the ground state represent non-fluorescent mechanisms.

1.5   Relationship between quantum yield, intrinsic lifetime and measured lifetime

For convenience, we can combine all the non-fluorescent processes that compete for deactivation of excited state electrons (e.g. internal

monoexponential lifetime decay

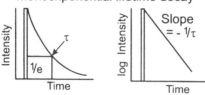

multiexponential lifetime decay

Figure 1.8. Mono- and multiexponential decays of the fluorescent lifetime. Monoexponential lifetime decays are plotted both on a linear (top left) and logarithmic (top right) scale.

conversion, k_i, phosphorescence, k_p and quenching, k_q) into a single rate constant, k_{nr} (non-radiative). As we have seen before (at steady state):

$$Q = \frac{\text{photons emitted}}{\text{photons absorbed}} = \frac{k_f}{k_f + k_{nr}},$$

where k_f is the rate constant for the decay of fluorescence. The intrinsic lifetime, τ_o, is defined as the lifetime of the excited state in the absence of any processes that compete for deactivation of the excited state electrons, and is therefore the inverse of the rate constant for the decay of fluorescence:

$$\tau_o = \frac{1}{k_f}.$$

In the real world, non-radiative processes occur which shorten the lifetime of the excited state so that the measured lifetime, τ_F, is a combination of the intrinsic lifetime, τ_o, and the non-fluorescent processes that compete for deactivation of excited state electrons, k_{nr}. Since

$$\tau_F = \frac{1}{\left(k_f + k_{nr}\right)},$$

$$Q = \frac{\text{photons emitted}}{\text{photons absorbed}} = \frac{k_f}{k_f + k_{nr}} = \tau_F / \tau_o.$$

τ_F is always less than τ_o and, therefore, τ_F/τ_o never exceeds 1. Because $Q = \tau_F/\tau_o$ and τ_o can be calculated, τ_F is a measure of the quantum yield.

1.6   Quenching

Various competing effects can cause non-radiative relaxation of excited state electrons to the ground state. Since these effects occur without emission of a photon, they are termed quenching. For example, collision of the excited fluorophore with another molecule resulting in non-radiative transfer of the excited state energy to this molecule is termed internal conversion or collisional or dynamic quenching. Dynamic fluorescence quenching can reflect the presence of proximate quencher molecules or groups which, via diffusion or conformational change, may collide with the fluorophore during the excited state lifetime. Such quenching acts to reduce the excited state lifetime, τ, and the quantum yield, Q. Due to collisional quenching the actual excited state lifetime is always less than τ_0 and Q is less than 1 (*Figure 1.9*).

A second type of quenching phenomenon, termed static or complex quenching, refers to static complexes of quencher and fluorophore which serve to reduce absorption (by depleting the population of fluorophores that can be excited) and hence reduce fluorescence output without changing the excited state lifetime. An excited fluorophore can also be quenched by resonance energy transfer if it is close enough to an 'acceptor' molecule to which its excited state energy can be transferred non-radiatively (see below).

1.7   Fluorophore–solvent interactions

The reason fluorescence is said to be exquisitely sensitive is that it reports on interactions with its environment which occur during the excited state lifetime. The excited state molecule can be considered a different molecule than that which exists in the ground state, and thus will display different properties in the excited state relative to the ground state. Fluorescent molecules absorb energy through dipoles; the ground and excited states have different dipole moments due to differences in the energy levels of their electrons. Solvent molecules which surround the fluorophore also have dipole moments that can interact with the dipoles of the excited state fluorophore. During the excited state life-

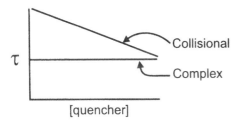

Figure 1.9. Effect of collisional and complex quenching on the fluorescent lifetime. Reproduced by permission of Marcel Dekker, New York, from Pesce, A.J., Rosen, C.-J. and Pashby, T. (1971) *Fluorescence Microscopy: An Introduction for Biology and Medicine*, p. 53.

time, the orientation of the solvent molecules is altered via these dipole–dipole interactions and they reach an equilibrium excited state of lower energy (Frank Condon excited state; *Figure 1.10*).

This has the effect of diminishing the energy separation between the ground and excited states (lower energy) resulting in a red shift in the emitted fluorescence. Such would be the case if the probe went from a neutral to polar solvent. Similarly, solvent–fluorophore interactions can occur where the energy separation between the ground and excited states is increased, resulting in a blue shift in the emission (*Figure 1.11*).

A fluorescent molecule in its excited state can be affected by a number of solvent (environmental) factors. For example, in addition to the polarity of the solvent, the temperature (by increasing collisional quenching or altering Boltzmann's distribution of molecules in the various vibrational states in the ground singlet state), pH (resulting in a different pK of the excited vs. ground state) and viscosity (by inhibiting solvent reorientation) can affect fluorophore–solvent interactions.

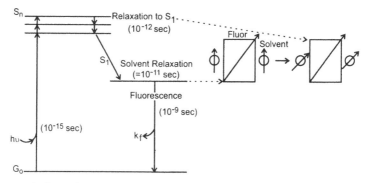

Figure 1.10. Frank Condon excited state. Following excitation of the fluorophore, solvent and fluorophore dipole moments can interact with each other, causing reorientation of dipole moments and altering the spectral properties of the fluorophore.

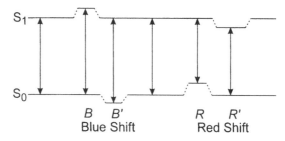

Figure 1.11. Blue vs. red shift in the spectrum of a fluorophore. A shift towards the blue portion of the electromagnetic spectrum results in the fluorophore possessing more total excitation energy, and can occur by either a lowering of the energy of the ground state (S_0) or an increase in the level of energy of the excited state (S_1). Conversely, a shift towards the red portions of the electromagnetic spectrum results in the fluorophore possessing less energy and can occur by either an increase of the energy of the ground state (S_0) or a decrease in the level of energy of the excited state (S_1). Reproduced by permission of Marcel Dekker, New York, from Pesce, A.J., Rosen, C.-J. and Pashby, T. *Fluorescence Microscopy: An Introduction for Biology and Medicine,* p. 67.

1.8   Polarization

Polarization is defined as the direction of oscillation of the electromagnetic wave that makes up light (*Figure 1.1*). Fluorophores preferentially absorb photons whose electron vectors are aligned parallel to the absorption transition dipole moments of the fluorophore (*Figure 1.12*). The transition dipole moments for absorption and emission have a fixed orientation within each fluorophore and are separated by some angle θ with respect to one another.

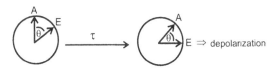

Figure 1.12. Fluorescence depolarization. Each fluorophore has fixed within its molecular framework an absorption (A) and emission (E) dipole moment. Fluorophores absorb light (energy) maximally when the light travels in space in a plane parallel to the plane of the absorption dipole moment, and emit light maximally in a plane oriented parallel to the emission dipole moment. These two dipole moments are separated from each other by an angle θ. During the excited state lifetime, τ, rotation of the fluorophore will cause further depolarization to take place. The amount of depolarization is related to the rigidity of the fluorophore environment.

Using polarized light, it is possible to selectively excite those fluorophore molecules whose absorption transition dipoles are parallel to the electron vector of the exciting light. The angle between the transition dipole moments for absorption and emission determines the maximum (limiting) measured polarization (p_0) or anisotropy ($r_0 = [(1/p - 1/3)^{-1}]$). Rotational motion (further displacement of the emission dipole from its starting position) which occurs during the lifetime of the excited state, will further lower the observed emission anisotropy. The rate of rotation is dependent both on the size of the fluorophore (or the macromolecule to which it is bound) and the viscosity of its immediate environment. The Perrin equation relates the fluorescence anisotropy (polarization) of a solution to the rotational mobility of the fluorescent molecule (either free fluorophore or fluorophore attached to a macromolecule):

$$r_0/r = 1 + RT\tau/\eta V,$$

where r_0 is the limiting anisotropy (the anisotropy observed in the absence of any rotation of the fluorophore), r is the measured anisotropy, η is the 'viscosity' of the environment, V is the volume of the rotating species, R is the universal gas constant, T is the temperature in degrees Kelvin and τ is the lifetime of the excited state of the fluorophore. The time-dependent decay of anisotropy for a spherical molecule is a single exponential such that:

$$r(t) = r_0\, e^{-t/\phi}$$
$$r_0/r = 1 + \tau/\phi,$$

where ϕ is the rotational correlation time of the fluorophore and is equal to:

$\phi = \eta V/RT$.

Polarization (anisotropy) is calculated by measuring the emission collected with a polarizer placed in the excitation light path and an analyser placed in the emission light path oriented either parallel or perpendicular to the excitation polarizer (*Figure 1.13*), and is defined as:

$p = (I_\perp - I_{||}) / (I_\perp + I_{||})$,

where $I_{||}$ is the fluorescence emission measured in the plane parallel to the plane of excitation and $I_\perp$ is the fluorescence emission measured in the plane perpendicular to the plane of the excitation. Polarization measurements provide information on the rotational mobility of proteins, from which their molecular weight and shape can be inferred, as can associations of proteins with each other, protein denaturation, the internal flexibility of proteins and a number of physical properties of membranes including viscosity, phase transitions, chemical composition and the effects of perturbation of membranes on these properties.

1.9   Fluorescence resonance energy transfer (FRET)

Fluorescence can be used as a spectroscopic ruler to study and quantify the interactions of cellular components with each other as well as con-

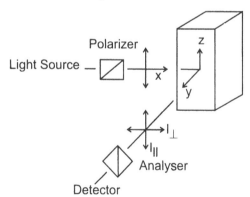

Figure 1.13. Instrumental approach for measuring fluorescence polarization. Plane-polarized excitation light is generated by passing light through a polarizer. This plane-polarized excitation light is absorbed maximally by fluorophores whose absorption dipole moments are oriented parallel to the plane of the excitation light. This results in only a certain sub-population of fluorophores being excited (an anisotropic collection of fluorophores). All of the excited fluorophores possess the same orientation of absorption and emission dipole moments. Fluorescence emitted from the fluorophore is measured both parallel ($I_{||}$) and perpendicular ($I_\perp$) to the plane of polarization of the exciting light. As the molecule rotates during its excited state lifetime, the amount of emission oriented parallel to the plane of excitation will decrease, while the amount of emission oriented perpendicular to the plane of excitation will increase.

formational changes within individual molecules at the molecular level (*Figure 1.14*). Resonance energy transfer (RET) is a process by which a fluorophore (donor, D) in an excited state may transfer its excitation energy to a neighbouring chromophore (acceptor, A) non-radiatively through dipole–dipole interactions (*Figure 1.14*).

In principle, if one has a donor molecule whose fluorescence emission spectrum overlaps the absorbance spectrum of a fluorescent acceptor molecule, they can exchange energy between one another through a non-radiative dipole–dipole interaction. This energy transfer manifests itself by both quenching of donor fluorescence in the presence of acceptor and increased (sensitized) emission of acceptor fluorescence (*Figure 1.15*).

Energy transfer efficiency varies most importantly as the inverse of the sixth power of the distance (radius, *r*) separating the donor and acceptor chromophores. The energy transfer to the acceptor is non-radiative (is not mediated by a photon) and requires the distance between the chromophores to be relatively close (usually not exceeding 10–100 Å). The phenomenon can be detected by exciting the labelled specimen with light of wavelength corresponding to the maximal absorption (excitation) of the donor and detecting light emitted at the wavelengths corresponding to the maximal emission of the acceptor, or by measuring the fluorescent lifetime of the donor in the presence and absence of the acceptor. The dependence of the energy transfer efficiency on the donor–acceptor

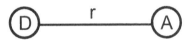

Figure 1.14. Principle of using fluorescence resonance energy transfer (FRET) as a spectroscopic ruler to measure the distance (radius) between a donor (D) and acceptor (A) fluorophore.

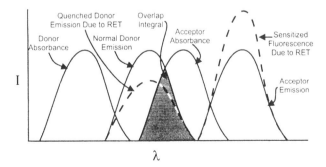

Figure 1.15. Measurement of FRET via changes in donor and acceptor emission intensities. Spectral overlap (hatched area) between donor emission and acceptor absorption is required for FRET. When this overlap occurs, and the donor and acceptor are separated by less than 100 Å, donor excitation energy can be transferred non-radiatively to the acceptor. This results in quenching of the donor fluorescence emission and an increase in the emission of the acceptor (sensitized emission).

separation provides the basis for the utility of this phenomenon in the study of cell component interactions. There are certain conditions that need to exist for RET to occur: (1) the donor must be fluorescent and of sufficiently long lifetime; (2) the transfer does not involve the actual reabsorption of light by the acceptor; and (3) the distance between the donor and acceptor chromophores needs to be relatively close (usually within 10–50 Å).

The rate of energy transfer (K_T) and the energy transfer efficiency (E_T) are both related to the lifetime of the donor in the presence or absence of the acceptor:

$$K_T = (1/\tau_D)\,(R_o/r)^6,$$

where R_o is the critical Förster distance, the donor–acceptor separation for which the probability of transfer equals that of the donor de-excitation rate in the absence of acceptor, τ_D the lifetime of the donor in the absence of the acceptor and r the distance separating the donor and acceptor molecules. R_o can be calculated from:

$$R_o = [8.75 \times 10^{-25}\,\eta^{-4}Q_o\,K^2 J(\lambda)]^{1/6},$$

where K^2 is the 'orientation factor', which describes the spatial relationship between the donor and acceptor absorption and emission dipoles (in 3D space) with respect to one another (*Figure 1.16*) (and $K^2 = 2/3$ if the orientation of the donor and acceptor are random), η is the refractive index of the medium, Q_o is the quantum yield of the donor in the absence of acceptor and $J(\lambda)$ the overlap integral.

E_T is related to r by:

$$r = R_o\,(1/E_T - 1)^6,$$

where

$$E_T = (\tau_{DA}/\tau_D)$$

and τ_{DA} is the lifetime of the donor in the presence of the acceptor. Thus, by measuring the lifetime of the donor with and without acceptor, one can calculate the distance between donor and acceptor.

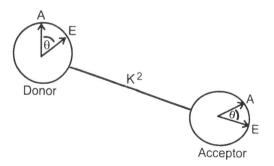

Figure 1.16. Orientation factor, K^2, is a measure of the angle (orientation) between the donor and acceptor absorption and emission dipole moments, and is given by the relationship $K^2 = \cos\phi_{DA} - 3\cos\phi_D\cos\phi_A$. Because of the usually rapid rotation of the donor and acceptor molecules, the statistical average of $K^2 = 2/3$ can be employed.

2 Fluorescence microscopy

2.1 Advantages of fluorescence microscopy

Fluorescence microscopy allows the user to look inside cells and tissues with exceptional detail. It provides the user with acquisition of spatial information (and in living specimens, temporal alterations in this spatial information) of objects that are either intrinsically fluorescent (primary or autofluorescence) or which have been coupled to extrinsic fluorescent molecules (secondary fluorescence), in samples too small to be seen with the naked eye. The combination of the specificity inherent in current fluorescence labelling techniques with the sensitivity of the fluorescence microscope has led to the ability to detect very small amounts of material with very high sensitivity and precision.

Fluorescence microscopy was developed in the early part of the 20th century. The basic function of the fluorescence microscope is to deliver excitation energy to the fluorescing species in the specimen and to separate the much weaker emitted fluorescence light from the brighter excitation light (*Figure 2.1*). In doing so, only the emitted light reaches the detector and a high contrast image is generated.

The use of fluorescence as a means of detecting objects through a microscope offers many advantages:

(i) *Specificity*. Fluorescence excitation and emission spectra are usually characteristic of a molecule's structure and composition. Different molecules usually display distinct excitation and emission spectra, which can be selectively used for the analysis of complex mixtures of molecular species.

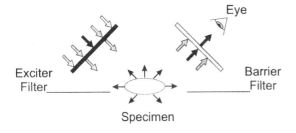

Figure 2.1. Principle of a fluorescence microscope. Adapted with permission of Olympus-America, New York, from Abramowitz, M. (1993) *Fluorescence Microscopy: The Essentials.*

15

(ii) *Sensitivity.* Detection of fluorescence can be carried out with only small numbers of fluorescent molecules. If discrimination of auto-fluorescence from specific fluorescence can be accomplished, highly sensitive fluorescence detection is possible.

(iii) *Quantitation.* Fluorescence provides quantitative measurements because the emitted fluorescence is directly related to the fluorescence quantum yield of the fluorophore. Quantification is feasible at relatively low concentrations because of the greater sensitivity associated with emission as opposed to absorption processes.

(iv) *Environmental sensitivity.* Fluorescence is extremely sensitive to the immediate physical and chemical environment; fluorophores have been designed for sensing environmental factors (pH, viscosity etc.) and other related parameters (e.g. Ca^{2+}).

(v) *High temporal resolution.* Fluorescence measurements can be used to detect very fast chemical and molecular changes in specimens.

(vi) *High spatial resolution.* Fluorescence can be measured from single molecules if the molecules contain a sufficient number of fluorophores (approx. 10–50 per μm^3). Cellular components with dimensions below the diffraction-limited resolution of the light microscope and their interactions can be visualized by using fluorescence resonance energy transfer techniques.

2.2   Fundamental optics of microscopy

The job of the fluorescence microscope is to collect as much of the emitted fluorescence given off by the object as possible while allowing visualization of fine detail. To understand how this is accomplished, it is necessary to understand the concepts of numerical aperture, resolution and magnification. Numerical aperture (NA) is a measure of the efficiency of the microscope objective with respect to the collection of emitted fluorescence, while magnification defines the size and resolution defines the detail of an object that can be observed through the microscope. Magnification and resolution of objects in the microscope enable the observer to see objects too small to be seen with the naked eye.

2.3   Numerical aperture

Because emitted fluorescence of microscopic specimens is usually relatively weak, it is very important to collect as much of the emitted fluorescence as possible. In fluorescence microscopes, the objective lens is responsible for collecting the emitted fluorescence from objects. As illustrated in *Figure 2.1*, fluorescence emission from a specimen occurs spherically in all directions regardless of the direction of the excitation energy. If we consider the object as a point source, we can define an angle, α, as the angle (direction) of the emitted fluorescence given off from the object relative to the original direction of the exciting light (*Figure 2.2a*).

It should be apparent that objectives that are capable of collecting light over a large angle α will provide brighter images. Such objectives are said to possess large NA where NA = $\eta\sin\alpha$, η being the refractive index of the medium between the specimen and the objective. The theoretical NA in a dry objective in air is 1.0, which is equivalent to a full 180° aperture angle (2α) of the objective. In practice however, it is not possible to obtain an NA of 1.0 (although one can obtain NA values of 0.95 in high dry magnification objectives), due to the fact that the emitted light at larger α is refracted more due to mismatches between the refractive indices, η, of the medium between the front surface of the objective and the coverslip (*Figure 2.2a*). To increase the NA of an objective, it is necessary to match more appropriately the refractive indices between the front surface of the objective and coverglass. Since the η of glass is 1.515, placing immersion oil with an η of 1.515 between the front lens of the objective and the coverslip effectively eliminates all reflections on the path from the object to the objective (*Figure 2.2b*), increasing the NA of the objective and hence the amount of light that can be captured by the objective.

2.4 Resolution

Resolution is defined as the smallest distance between two small objects that can still be discerned as two separate objects. The image formed in

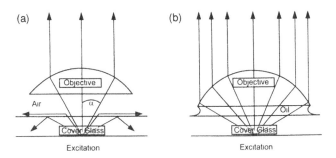

Figure 2.2. Influence of the refractive index of the immersion medium [(a) air; (b) oil] on the theoretical maximum of the numerical aperture.

the intermediate image plane of an objective lens of an infinitely small point source of light is the diffraction image of the original object (*Figure 2.3*). The centre of this diffraction image is known as the Airy disc and its radius (r_{Airy}) is given by:

$$r_{Airy} = 0.61(\lambda/NA),$$

where λ is the wavelength of light and NA the numerical aperture. When two small objects separated by some distance d are imaged, these two small objects can be distinguished until this distance $d = r_{Airy}$ (the centre-to-centre distance between the two objects equals the radius of the object). At $d = r_{Airy}$, the diffraction images of the two objects will still just be distinguishable because there is a 26.5% dip in intensity between the two diffraction images, even though they are overlapping (the Rayleigh criterion). If the objects' images further overlap and d becomes less than the Rayleigh criterion, the two small objects cannot be distinguished as individual objects. Eventually, the two objects reach a point where there is no intensity difference between them (at $d = 0.78\ r_{Airy}$ – the Sparrow criterion). Because the diameter of the Airy disc is governed by the objective NA and the λ of the light, the higher the NA of the objective, the smaller d will be. A shorter wavelength is also beneficial for increasing objective resolving power. *Table 2.1* shows the values for the limit of resolution for some objectives ($\lambda = 550$ nm). As can be seen from the table, use of higher NA objectives provides better resolution.

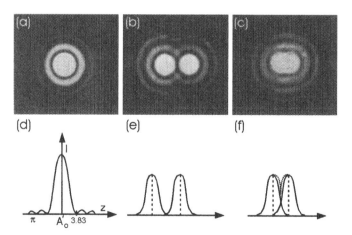

Figure 2.3. Airy disc and intensity distribution in the diffraction pattern. (a, d) Airy disc and intensity distribution. (b, e) Two Airy discs and their intensity distributions. The separation of the two Airy discs is greater than their radii, and they are resolvable. (c, f) Two Airy discs and their intensity distributions under conditions where the centre-to-centre distance equals the radius (Rayleigh criterion). Adapted with permission of Plenum Press, New York, from Inoué, S. (1986) *Video Microscopy.*

Table 2.1. Limiting resolution of microscope objectives

Objective/NA	d (µm)
5×/0.15	2.2
10×/0.30	1.1
20×/0.50	0.7
40×	0.4
40×/1.30 oil	0.3
63×/1.40 oil	0.2
100×/1.30 oil	0.3

Adapted from Kapitza, H.G. (1996) *Microscopy From the Very Beginning.* Carl Zeiss, Jena.

2.5 Magnification

Magnification is defined as the apparent enlargement of an object by an optical instrument (i.e. the power to cause objects to appear larger than they are). The overall magnification of a microscope should be higher than 500 times, but less than 1000 times, the objective aperture. This will then be the range of the useful magnification.

The transmittance of light intensity (intensity transmitted as a percentage of incident intensity) through the optical components of a microscope is critical in fluorescence microscopy, where high resolution imaging requires high magnification with minimal loss of image brightness. Therefore, one should employ as high transmittance objectives as possible. Due to the relationship between NA and magnification, the overall brightness of the image decreases rapidly as the magnification increases. In the typical fluorescence microscope in use today, which employs incident or epi-illumination (see below), the objective lens serves both as a condenser and objective. Hence, a high aperture objective will increase the signal as the light concentrating power of the objective (serving as a condenser) is proportional to $(NA)^2$ and the light gathering power of the objective is also proportional to $(NA)^2$. Thus, at least up to everything but the highest NA objectives, brightness varies as $(NA)^4$. However, brightness varies as $1/(Mag)^2$ and, therefore, brightness is proportional to $(NA)^4/(Mag)^2$.

The eyepieces through which the individual observes the image further magnify the image, decreasing the overall observed intensity of the image (the brightness of the image will be inversely proportional to the square of the eyepiece magnification). Therefore, one should use as low a power eyepiece as is consistent with convenient observation. Thus, for extremely low light level situations, such as fluorescence, one can maximize image brightness by using the highest NA objective and the lowest overall magnification. This leads to a compromise between attempts to increase the resolution and to maintain a high signal-to-noise ratio, which decreases as the image becomes dimmer. As many as 12 lens elements are used in an objective and these lenses are usually coated to avoid reflection losses. Coating all of the lens surfaces increases the transmittance of the objective by 85%. Each uncoated glass element in a lens causes roughly a 4% loss of transmittance at all wavelengths. Therefore, coating of lenses is essential to improve the transmission and decrease the flare due to multiple reflection at the glass surfaces.

2.6 Optical paths of fluorescence microscopes

Two different types of optical paths have been employed in fluorescence microscopy. Historically, transmitted light illumination (diascopic fluorescence) had been used for fluorescence microscopy (*Figure 2.4a*). In this type of illumination path, a condenser focuses the exciting light on to the microscope field. Successful use of this optical arrangement requires very careful alignment of the condenser and objective lenses, and because a bright-field condenser is used for illumination, most of the exciting light enters the objective. This requires very high quality filters capable of separating excitation light from fluorescent emission, which in practice is difficult to achieve. The use of high numerical aperture oil dark-field condensers was adapted to this illumination arrangement, as this design allowed excitation light to be delivered to the specimen at steep angles and prevented most of the excitation light from entering the objective (*Figure 2.4b*). While this helped, this illumination design is still limited in that it cannot be used for simultaneous fluorescence/

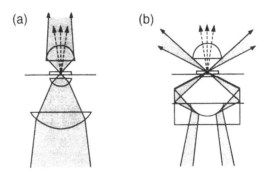

Figure 2.4. Comparison of transmitted light (diascopic) bright-field fluorescence (a) and transmitted light oil dark-field fluorescence (b) condensers. It is difficult to separate excitation and fluorescence emission using a transmitted light bright-field fluorescence condenser because both excitation and emission enter the objective. In contrast, the high NA oil dark-field condenser sends light to excite the specimen at steep angles, and thus very little excitation light enters the objective (b). Adapted with permission of Olympus-America, New York, from Abramowitz, M. (1993) *Fluorescence Microscopy: The Essentials.*

phase or fluorescence/differential interference contrast imaging. The design of the objective lens required for transmitted fluorescence illumination is also such that the NA is reduced relative to a normal lens.

By far the most commonly employed illumination system is that invented by Ploem and termed Ploem, incident-light or epi-illumination (*Figure 2.5*). In epi-illumination, the exciting light is reflected into the back aperture of the objective (which acts as a condenser) by a dichromatic beam-splitting mirror. Fluorescence is collected by the objective and light forming the image passes *through* the dichromatic beam-splitting mirror to either the eyepieces or a camera. Because the wavelength of emission is of lower energy (longer wavelength) than that of the excitation (Stokes' law), the dichromatic beam-splitting mirror is designed so that it will reflect light of shorter wavelength (the excitation) but will transmit light of longer wavelengths (the emission). The dichromatic beam-splitting mirror reflects any initial or refracted exciting light, allowing only the emitted fluorescence light to pass through and reach the detector.

Figure 2.6 illustrates an example of the separation of blue excitation light from green fluorescence (this would be the arrangement for use with the fluorophore fluorescein). A light source (see below) outputs excitation energy of varying wavelengths; the appropriate wavelength for absorption of excitation energy by the fluorescein molecule is selected through the use of a filter that transmits light below 490 nm (SP490), wavelengths which correspond to the maximal absorbance of fluorescein. Note that other wavelengths of light also pass through the filter (i.e. red) although to a much lesser extent. The 490 nm excitation light reaches the dichromatic mirror (CBS 510 nm) where it is reflected into the back aperture of the objective and focused on to the specimen. As we previously discussed, the dichromatic beam-splitting mirror has the

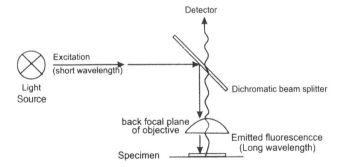

Figure 2.5. Reflected light vertical illumination system (epi-illumination) commonly used for fluorescence microscopy. In this arrangement, the objective acts as both a condenser and objective, excitation light moves away from the objective during fluorophore excitation, the area illuminated is that which is observed and the full NA of the objective is usable. In addition, the presence of a dichromatic beam splitter efficiently reflects excitation energy on to the specimen and allows only long wavelength emitted fluorescence to reach the detector.

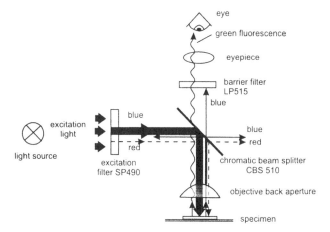

Figure 2.6. Hypothetical filter combination for observation of fluorescein: separation of fluorescence excitation light from fluorescence emission light by a dichromatic beam splitter (CBS 510). Blue excitation light of 490 nm is generated when the excitation light passes through the short-pass (SP490) filter. This filter allows light below 490 nm to pass through the filter unattenuated. The filter does not have an absolute cut-off at 490 nm so a small amount of light of wavelengths above 490 nm also passes through the filter. The excitation light then hits the CBS 510 nm and light below 510 nm is reflected downward through the objective and on to the sample. The emitted fluorescence and scattered excitation light then re-enters the objective and hits the CBS 510 nm. This CBS allows light above 510 nm (fluorescence emission) to pass through the mirror, while reflecting light below 510 nm, preventing this light from reaching the detector. This effectively separates the excitation light (which is substantially larger in amount) from the much smaller amount of emitted fluorescence. As is the case with the excitation filter, the CBS does not possess an absolute cut-off at 510 nm and thus a small fraction of the blue excitation light leaks through the CBS. A long-pass (LP 515) barrier filter, which allows only light of wavelengths of at least 515 nm to pass through the filter, removes the residual excitation light.

property that it will reflect light of shorter wavelengths but will transmit light of longer wavelengths. In this case, the CBS 510 nm denotes that the dichromatic beam splitter will reflect light below 510 nm into the objective, and let light of wavelengths longer than 510 nm pass through the dichromatic beam-splitting mirror without reflection. Note that the reflection of the excitation light is not 100%; a small amount of the 490 nm light passes through the dichromatic beam splitter without being reflected. In addition, not all of the light of wavelengths greater than 510 nm (i.e. green and red) passes through the beam splitter; a small amount of this light is also reflected on to the specimen. The emitted fluorescence (green light) is collected by the objective and passes through the 510 nm dichromatic reflector without reflection and then through another 'barrier' filter (LP515), which in this case allows light of wavelengths greater than 515 nm to reach the detector. This last filter effectively prevents any scattered non-reflected excitation light from reaching the detector.

2.7   Components of the fluorescence microscope

The main components of a fluorescence microscope are: (1) excitation light sources; (2) wavelength selection devices; (3) objectives; (4) detectors; and (5) stages and specimen chambers.

2.7.1   *Excitation light sources*

Light sources for fluorescence microscopy consist of tungsten or halogen lamps for transmitted or incident illumination and mercury, xenon (e.g. Hg, Xe or Hg/Xe combination) or metal halide arc lamps or lasers for epi-illumination (*Figure 2.7*). Selection of fluorescence light sources depends on the fluorescent probes being used (e.g. maximal wavelengths of absorption and quantum yield).

Tungsten and halogen (filament lamps) have not been used with a great deal of success in fluorescence microscopy owing to the fact that the majority of their emission occurs in the red or infrared portion of the light spectrum, and many of the available fluorophores require excitation in the blue/green portion of the visible light spectrum. Thus, in epi-illumination fluorescence microscopes, Hg, Xe or Hg/Xe combination lamps and lasers have become the favoured light sources. Each has benefits depending on the wavelength and strength of excitation energy required in a given experimental situation.

The design of the Hg and Xe light sources is similar except for the gas enclosed in the lamp. A Hg light source consists of two electrodes sealed under high pressure in a quartz glass bulb which also contains Hg. When the lamp power supply is turned on, it sends high voltage pulses to the electrodes, which ionizes some of the gas in the bulb, igniting the lamp. Subsequently, these ions, under low voltage, carry the current which generates the light in-between the two electrodes. The lamp gets very hot vaporizing the Hg, creating a huge amount of pressure inside the glass bulb. When the lamp gets very hot it gives off high intensity light concentrated at certain wavelengths (mercury lines). Hg light sources emit energy at discrete wavelengths (e.g. 365, 400, 440, 546 and 580 nm) making this light source useful for caged compound release, Hoescht, 4′,6-diamidino-2-phenylindole (DAPI), fluorescein and rhodamine-based fluorophores. For example, measurement of intracellular pH using 3′-*O*-acetyl-2′,7′-bis(carboxyethyl)-4- or -5-carboxyfluorescein (BCECF) works well with a Hg arc lamp as the excitation source. Xe lamps have a spectrally uniform intensity profile from the ultraviolet to the far red; this light source is particularly suited for use of fura-2, a calcium-sensitive dye. The metal halide lamp spectral output is the same as the Hg arc lamp but the energy per unit area for metal halide is less

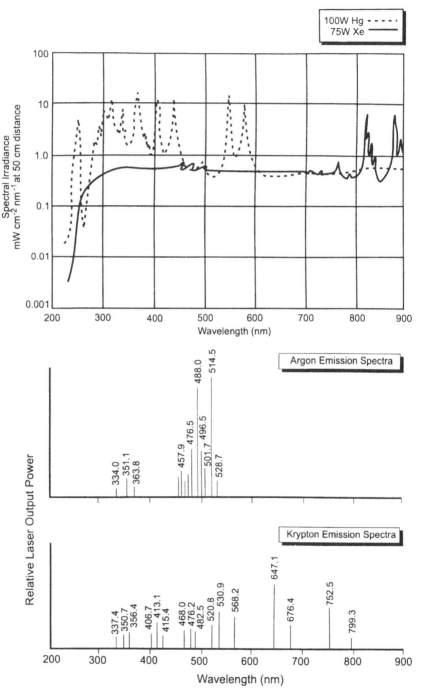

Figure 2.7. Emission spectra of lamps and lasers commonly used for fluorescence microscopy. Note that the output of a Xe lamp is relatively constant in the visible wavelength range, while Hg lamps put out high light intensity at specific wavelengths in the visible spectrum. Lasers put out monochromatic light whose wavelengths and intensities vary depending on the type of laser. Adapted with permission of Academic Press, San Diego, from Spring, K.R. and Long, R.J. (1989) *Meth. Cell Biol.* **29**: 269–289.

than the Hg arc lamp. However, the 150 W metal halide bulb has an average life-span of 1000 hr compared to 200 hr for a 100 W Hg and 400 hr for a 75 W Xe bulb. The tendency for arc lamps to flicker or wander can be minimized by an initial 'burn-in' period, and it is advisable to use as small an arc size as possible.

Lasers (Light Amplification by Stimulated Emission of Radiation) represent another option for providing excitation energy for fluorescence microscopy. Lasers work on the same principle that we discussed in Chapter 1 regarding fluorescence. However, unlike the case in fluorescence, in a laser the goal is to generate a 'population inversion' where most of the electrons are excited at the same time and reside in the same excited energy state until some external disturbance comes along to dislodge it, whereupon the excited electrons fall to their lower energy state with an accompanying pulse of light. In a laser, this external disturbance is a photon passing nearby. The disturbing photon stimulates the excited atom to emit its energy as a photon, and this new photon of light moves off in the same direction as the disturbing photon with both of their waves precisely aligned. This 'stimulated emission' is the key to the operation of the laser. These two photons can stimulate other excited atoms to give off their energy in the same way and, like an avalanche, the emitted light is amplified. Lasers offer monochromatic light of very high intensities and can be used in continuous wave (CW) or pulsed modes of operation. Pulsed lasers are finding more and more applications in fluorescence microscopy, especially in two-photon excitation microscopy (TPEM) and in the two- and three-dimensional measurement of fluorescent lifetimes in intact cells and tissues.

An important concern with the use of lasers in microscopy is the potential for scattering and diffraction due to dirt and interference on optical surfaces in the beam path. This can be overcome by using a multimode fibre-optic light pipe. Numerous types of lasers exist, each of which has its own characteristic spectral and power output (*Figure 2.7*).

2.7.2   *Wavelength selection devices*

The human eye is able to detect light between the wavelengths of 400 and 700 nm. Light of wavelengths 340–400 nm is considered to be near ultraviolet and essentially invisible to the human eye. The portion of the electromagnetic spectrum visible to the human eye is as follows: 400–430 nm, violet; 430–500, blue; 500–560, green; 560–620, yellow to orange; and 620–700 nm, orange to red. Light above 700 nm is considered near infrared and is also invisible to the human eye.

Since the fluorescence microscope performs basically as a filter fluorometer, a variety of wavelength selection devices have been developed to select the wavelengths of excitation and emission. These include tunable lasers, monochromators, individual versus multiple band-pass filters, acousto-optical tunable filters (AOTF) and liquid crystal tunable

filters (LCTF). Filters are used to select a certain part of the electromagnetic spectrum for transmission, while preventing the rest of the
electromagnetic spectrum from passing through the filter. In the fluorescence microscope, excitation light of a specific energy level (wavelength) from a Hg or Xe arc lamp or a laser is isolated using an exciter
filter, reflected on to the specimen by a dichromatic beam-splitting mirror, and the emitted fluorescence is viewed through a cut-off or barrier
filter which blocks all wavelengths below a set wavelength from reaching the observer (*Figure 2.6*).

The simplest filters consist of colour-filtered or absorption glass
(including gelatin), which attenuate light solely by absorption. The
amount of attenuation is dependent on the thickness of the glass. While
this type of filter is long lasting and inexpensive, its usefulness is limited by high autofluorescence, low peak transmittance and poor discrimination of closely spaced wavelengths.

Thin-film coating of either a metallic nature (for fully reflective mirrors and neutral density filters) or interference coatings (for interference
filters) are the most widely used type of wavelength selection devices.
Advantages of thin-film coating are that almost any filter type can be
designed (including multiple band-pass filters; see below) and that the
desired wavelength, bandwidth and blocking level (amount and wavelengths of light prevented from passing through the filter) can be controlled with high precision. Drawbacks of this type of filter include a
finite blocking wavelength range, the materials used for coating are limited in their range of transparency, and the interference coatings are
sensitive to the angle of incidence of the light. As the angle of incidence
increases, the filter characteristics shift to shorter wavelengths. Several
standard exciter filter–dichromatic beam-splitting mirror–barrier filter
combinations are available from each microscope manufacturer. The
choice of which combination to employ will depend on the fluo-
rochrome(s) employed. Broad-band filters provide more signal but less
contrast (i.e. the ability to distinguish specific vs. non-specific fluorescence or the ability to distinguish individual fluorophores when more
than one fluorophore is present), while narrow-band exciters provide
less signal but more contrast.

The various types of filters are described in *Figure 2.8*. A short-pass
filter will pass light up to a certain wavelength, but not allow light of
higher wavelengths to pass through the filter (i.e. in *Figure 2.6*, the excitation SP490 filter would allow passage of wavelengths below 490 nm
through the filter but not wavelengths over 490 nm). In contrast, a long-
pass filter will allow only wavelengths above a certain wavelength to
pass through, and will prevent light of lower wavelengths from passing
through the filter (i.e. in *Figure 2.6*, the LP515 allows only light of wavelengths greater than 515 nm to pass through the filter). Both short-pass
and long-pass filters are named by the wavelength at which the transmission of light is 50% of maximum.

Dichromatic beam-splitting mirrors reflect the short wavelength

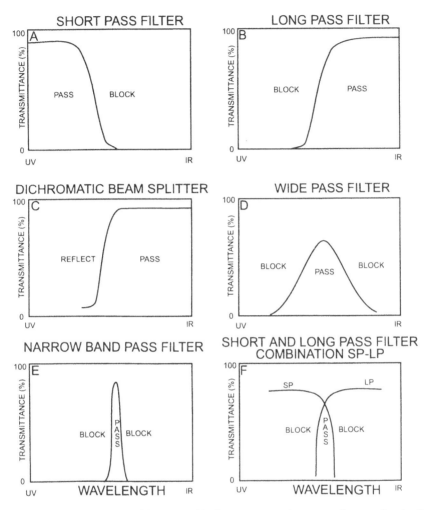

Figure 2.8. Various types of filters used in fluorescence microscopy. See text for details. Adapted with permission of Olympus-America, New York, from Abramowitz, M. (1993) *Fluorescence Microscopy: The Essentials.*

excitation light to the back aperture of the objective, allow the long wavelength emitted fluorescence to pass to the detector, but prevent the passage of the shorter excitation wavelengths to the detector. Optimally, greater than 90% reflectance of the excitation light and 90% transmission of the emission light is desired. Dichromatic beam splitters can suffer from polarization-dependent transmission of light. The P plane (also known as the TM mode, standing for transverse magnetic) is the component of the light beam's electric field that is parallel to the plane of incidence of the beam splitter, and the S plane (also known as the TE mode, standing for transverse electronic) is the component of the light beam's electrical field that is perpendicular to the plane of incidence of the dichromatic beam splitter. While filter manufacturers attempt to minimize the polarization dependence of transmission if possible, it can have an advantage. Because the S and P plane transmission are spectrally distinct, it is possible to perform simultaneous fluorescence and differential interference contrast microscopy without any light loss (see Chapter 4).

In addition to filters which have specific cut-off wavelengths of light, above or below which they will not pass, it is also possible to construct filters that can transmit only a specific range of wavelengths (interference or band-pass filters). For example, both wide-band-pass and narrow-band-pass filters exist, which allow wavelengths of only a certain range to pass through (*Figure 2.8d* and *e*). The centre wavelength of a band-pass interference filter depends on the angle of incidence of illumination. Excitation and emission band-pass interference filters are normally operated at 0° angle of incidence, while dichromatic mirrors are operated at 45° angle of incidence. Dichromatic beam-splitting mirrors can be ordinary short- or long-pass filters or even band-pass filters inclined at an angle. The narrower the range of wavelengths that can pass through the filter, the more specific is its ability to discriminate individual fluorophores in a mixture. Of course with specificity comes a price, and that is less intensity. These filters are usually named by their centre wavelength (the arithmetic means of the wavelengths at 50% of peak transmission) and by the range of wavelengths (bandwidth) of light they transmit at 50% peak transmittance (full width half maximum, FWHM). For example, BP 546/30 is an interference filter with maximal transmission at 546 nm (for rhodamine excitation) and transmits light from 531 to 561 nm. Sandwiching an SP and LP filter together also has been used to generate specific band-pass filters (*Figure 2.8f*). A typical combination of filters for the excitation of fluorescein and rhodamine is shown in *Figure 2.9*.

Historically, when examining a single specimen stained with multiple fluorophores, the use of separate filters for each fluorochrome was often

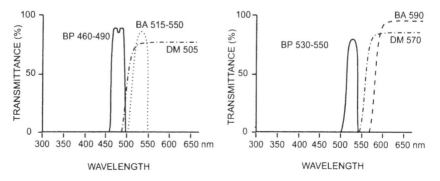

Figure 2.9. Common filter set for viewing fluorescein-labelled (left) and rhodamine-labelled (right) structures. The use of a BA 515–550 nm interference filter allows separation of fluorescein and rhodamine signals. Adapted with permission of Olympus-America, New York, from Abramowitz, M. (1993) *Fluorescence Microscopy: The Essentials.*

associated with image registration shifts due to varying amounts of wedge and tilt in the emission filter and dichromatic beam splitter, variations in the thickness and alignment of the different dichromatic beam splitters and mechanical vibration due to movement of the filters. One solution to this problem was to design single filters with multiple band-passes that can be used to image more than one fluorophore (i.e. double or triple band-pass filters, *Figure 2.10*). In newer microscopes image registration problems have been largely eliminated, so the need for multiple band-pass filters is decreasing.

In addition to employing light sources and filters as wavelength selection devices, two other types of wavelength selection devices are employed in fluorescence microscopy. Monochromators have been used because of the ease of wavelength selection and potential for wavelength scanning. However, monochromators suffer from being mechanically complex, large and expensive, and can generate spurious harmonic wavelengths. The other type of device used for wavelength selection is known as an electro-optic wavelength selection device and includes the AOTF and LCTF. In an AOTF, high frequency vibrations (50–200 MHz) from a piezo-electric device are used to generate compression waves in different types of crystalline material (*Figure 2.11a*). This leads to the generation of a diffraction grating due to the fact that the refractive index of the crystalline material increases as it is compressed. Advantages of this device include that it has no moving parts, no instability in intensity throughput, no wavelength drift, an efficiency of diffraction of the first order wave of approximately 85% (which provides the excitation wavelength) and wavelength switching times with a temporal resolution of 2–3 msec.

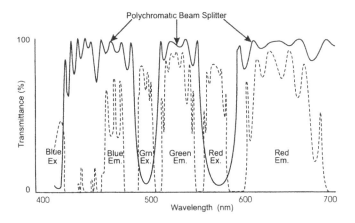

Figure 2.10. Spectra of a triple-band filter set designed for the dyes DAPI, FITC and Texas red. Adapted with permission of Chroma Technology Corp. from Reichman, J. *Handbook of Optical Filters for Fluorescence Microscopy*, P/N 61002 (1994).

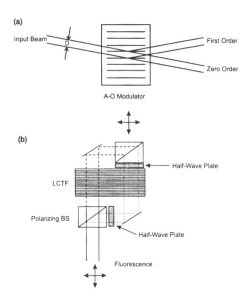

Figure 2.11. (a) Schematic representation of an acousto-optic tunable filter (AOTF). The input beam consists of white light which enters a TeO$_2$ crystal in which a travelling acoustic wave (vibrations) has been induced by a variable radio frequency source. This is similar to a bulk transmission diffraction grating; by varying the frequency it is possible to diffract out any monochromatic wavelength of light in a polarized manner with high precision. (b) Schematic representation of a liquid crystal tunable filter (LCTF). Liquid crystal are polymers whose orientation is voltage sensitive. By trapping a layer of liquid crystals between conductive plates and varying the voltage across the gap, the degree of retardation of polarized light can be altered in a controlled way. A stack of such variable retarders can be used as a wavelength selection device. Adapted with permission of the Microscopy Society of America, New York, from Chatton, J.Y. and Spring, K. (1993) *MSA Bull*, p. 329.

LCTFs are liquid crystals whose polymer orientation is voltage sensitive (*Figure 2.11b*). By varying the voltage in the liquid crystal medium, one can control the degree of retardation of polarized light. Stacking such variable retarders results in an interferometer which can perform wavelength selection. LCTFs are relatively new wavelength selection devices; their advantages include the lack of mechanical parts, 10–50 msec temporal resolution and transmission of approximately 30–50%. However, LCTFs display polarization sensitivity, an inability to transmit light below 400 nm, and the possibility of damage at high light flux levels. When a LCTF is placed between the excitation light source and the sample, care must be taken to protect the crystal from getting too hot (> 50°C) by using infrared (IR) rejection filters and a heat sink.

A discussion of filters for use in fluorescence microscopy would not be complete without mention of heat and neutral density filters. Heat filters are used to protect the excitation filters and detectors from infrared radiation. The heat filter is placed in front of the light source to reduce heat transfer to the excitation filter. These heat absorption filters can also reduce the UV excitation intensity, of potential benefit when performing live cell measurements. Neutral density filters reduce the intensity of all wavelengths of the exciting light equally. They may employ reflection and/or absorption. These filters are used for reducing the intensity of the exciting light to prevent photobleaching of chromophores and photodamage of living cells and tissue. The neutral density filter should be interposed between the heat filter and excitation filter to provide additional protection against damage to the interference filter coatings due to high temperatures generated by the light source. The use of a computer-controlled electronic shutter open only when experimental data is being collected, will also prolong filter life, as well as keeping exposure of the sample to harmful radiation at a minimum. There is another approach to prolonging filter life-span, and that is to use fibre-optic coupling of the light source to the microscope. This allows physical isolation of the heat of the excitation light source from the excitation wavelength selection device and provides uniform illumination as the fibre optically scrambles the light as it passes through the fibre. It is also possible to place the fibre in front of both the excitation light source and the filter changer, further isolating the microscope from mechanical vibration due to the mechanical filter switching device.

There are a number of methods for changing the wavelengths of excitation and emission while collecting data. One can manually move different filter sets into and out of the beam path of the microscope, but this usually results in vibrations and limitations in temporal resolution. Computer-controlled filter wheels, which hold up to 10 different interference filters and can be switched from one filter to another very rapidly, are now the most common means to mechanically alter excitation and emission wavelengths. Of course, it is also possible to construct systems with multiple light sources, detectors and wavelength-specific filters to further enhance the ability to image multiple fluorophores in the same object with high temporal resolution.

2.7.3 *Objectives*

A number of objectives exist for fluorescence microscopy. Objectives can be classified into transmitted-light and reflected-light versions. The first type is designed for use with coverslips (0.17 mm). Reflected-light objectives are signified by the prefix 'Epi-' and feature specially coated glass surfaces to avoid reflection in the optics ('antireflection coating'). More recently, infinity focused objectives have been developed. In infinity focused objective lenses, light rays emanating from a point in the object are parallel between the objective and the eyepiece. The advantage of infinity designed objectives is their insensitivity to other components introduced in the beam path, such as filters, analysers, compensators, differential interference contrast (DIC) prisms etc., which in finitely designed objective systems cause axial and lateral shifts of the light rays, making it difficult to get exactly overlapping images when performing double or triple fluorescence labelling experiments.

A partial list of objectives includes: (1) Ultra-Fluars (200–700 nm transmission); (2) Plan-Achromats (flat-field and colour corrected, usable at near UV wavelengths); (3) Plan-Neofluars (flat-field corrected with high transmission near UV); (4) Fluars, which have a wide spectral range; (5) Multi-immersion objectives, which can be used with or without a coverslip in a variety of different media; (6) Quartz objectives for use in the UV; (7) Epiplan-Achromats and Neofluars for use in reflected light; and (8) newly developed high numerical aperture water-immersion objectives (40 ×–60 ×, NA 1.2). These latter lenses are becoming increasingly popular, as they provide excellent images of living cells and tissues bathed in a medium, allowing fluorescence imaging of real-time cellular physiology.

The wavelength of light which will be transmitted through the objective, as well as the application for which it is being used, need to be considered when deciding on an appropriate objective. *Figure 2.12* and *Figure 2.13* illustrate the relative transmittance and depth of focus as a function of wavelength for a series of objective lenses. For example, Fluar lenses can be used at greater than 350 nm, while Quartz lenses can be used at less than 350 nm. Planapo lenses are excellent for photography but their many elements will diminish transmittance. The light throughput of modern objectives is only 15–50%. The selected objectives should also have negligible autofluorescence, no solarization (colour changes due to prolonged irradiation) and be picked to optimize brightness in the image.

Objective lenses can suffer from spherical aberration (light rays passing through different radii of a lens not coming to the same focus), astigmatism (variation in the focal length of the lens dependent on the incident plane of the light rays entering the lens), coma (the formation of comet-shaped patterns by off-axis beams), curvature of field, chromatic aberration (different colours of light are focused at different distances

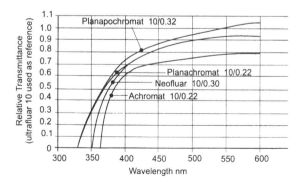

Figure 2.12. Relative transmittance as a function of wavelength of a variety of types of objective lenses. Adapted with permission of Springer-Verlag, Heidelberg, from Piller, H. (1977) *Microscope Photometry.*

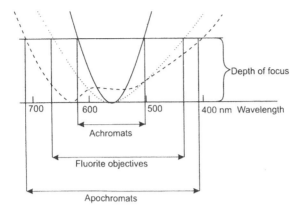

Figure 2.13. Depth of focus as a function of wavelength for various types of objective lenses. Adapted with permission of Springer-Verlag, Heidelberg, from Piller, H. (1977) *Microscope Photometry.*

from the lens), lens distortions and limited field of view. Objective lenses are mainly corrected for spherical and chromatic aberrations and flatness of the intermediate image. When these corrections are made, the objective can provide a very uniform image of high quality. *Table 2.2* lists the general types of objective lenses and their corrections for spherical and chromatic aberrations, as well as flatness of the intermediate image.

Different objectives have been designed for specific uses. Achromatic lenses (Achromat and Achrostigmat) are relatively inexpensive objectives, colour corrected for two wavelengths, which produce a flat image in the centre of the viewing field. The field of view is somewhat limited (up to 18 mm diameter) and these objectives are used for phase contrast microscopy. Plan and Epiplan (Achroplan, transmitted light; Achroplan Ph, phase contrast; Epiplan, reflected light (DIC); Epiplan HD, dark-field objective) are based on the Achromat objective but are improved in ways such that they are excellent lenses for photomicrography. These objectives feature good image contrast, flat fields of view (20–23 mm

Table 2.2.   Objective lens types and corrections

Lens type	Spherical	Chromatic	Flatness
Achromat	2 λs @ 2 NAs	Blue, red (broad range of visible spectrum)	No
F-Achromat	2 λs @ 2 NAs	Blue, red (broad range of visible spectrum)	Improved
Neofluar	Blue, green, red	Blue, green, red (full range of visible spectrum)	No
Plan-Neofluar	Blue, green, red	Blue, green, red (full range of visible spectrum)	Yes
Plan Apochromat	Dark blue, blue, green, red	Dark blue, blue, green , red	Yes

Adapted from Inoué, S. (1986) *Video Microscopy.* Plenum Press, New York.

diameter) and long working distances (the distance between the front element of the objective lens to the surface of the coverslip). Plan-Neofluar and Epiplan-Neofluar are excellent all-round objectives, spherically and chromatically corrected over a broad range of the visible spectrum with flat fields of view up to 25 mm diameter. A particular advantage of these lenses is that they transmit very well in the near UV range of wavelengths, and have found particular use when one wishes to maximize UV excitation at the 365 nm line of the Hg lamp. Plan Apochromat and Epiplan-Apochromat are the best and most costly of the 'standard' microscope objectives. They offer superb colour rendition and flat image fields of up to 25 mm diameter. These lenses are usually high NA and provide the highest resolving power. This requires use of oil immersion and therefore these lenses have smaller working distances.

In addition to these standard objectives, there exist special objectives such as water-immersion lenses, multi-immersion (oil, water, glycerine) objectives and lenses that transmit UV excitation. In biological applications involving living cells, photon efficiency of the objective is important. Increased photon efficiency can be attained by maximizing the NA. This can be accomplished by employing a minimum number of lens elements, maximizing the field of view and correcting any chromatic aberrations. Recently a series of objectives with a 200 mm focal length, a parfocal distance of 60 mm and a 25 mm image field has become available. This lens design results in objectives with longer working distances (allowing examination of thick specimens, microinjection and the use of thick-bottom chambers), higher NAs for both air and oil-immersion lenses (providing increased resolution and sensitivity), and very low (i.e. 0.5 ×) magnification lenses. At present, there is no independent, fully standard test procedure that exists to quantify the performance of an objective.

Some of the more important characteristics of the objective are inscribed on the objective lens barrel. These include the type of objective lens, its magnification, NA, the tube length, whether or not the objective should be used with oil and the thickness of the coverslip that is optimal for use with the specific objective. For example, Plan Apo 60 ×/1.40 oil 160/0.17 translates into Plan Apochromatic oil-immersion objective of 60 × magnification with a NA of 1.4, a tube length of 160 mm and which is to be used with a 0.17 mm thick coverslip. Selection of coverglass thickness is important, as this has to match the particular objective to avoid spherical aberrations. Some objectives have a correction collar to adjust the lens to the thickness of the coverglass or to the refractive index of the immersion medium. The magnification and immersion medium can also be indicated by coloured rings on the barrel of the objective. Coloured rings nearest the specimen identify the immersion medium, while coloured rings above the immersion coloured rings denote objective magnification. *Table 2.3* lists the colour-codes for objective immersion medium and magnification.

Table 2.3. Colour codes for objective immersion medium and magnification

Colour	Immersion medium	Magnification
Black	Oil	1, 1.25
Brown		2, 2.25
Red	Special	4, 5
Orange	Glycerol	
Yellow		10
Green		16, 20
Turquoise		25, 32
Light blue		40, 50
Cobalt (dark) blue		60, 63
White (cream)	Water	100

2.7.4 Detectors

Detectors for fluorescence microscopy allow the visualization of low levels of emitted fluorescence without photobleaching (and in the case of living specimens photodamage) to the specimen. They also allow real-time recording of living cell and tissue physiology. Detectors for fluorescence microscopy range from photomultiplier tubes, which are used for quantitative measurements, to imaging detectors, which allow the acquisition of two-dimensional images of fluorescence. Key considerations in choosing a detector are the required sensitivity, the level of noise inherent in the detector (and thus the signal-to-noise of the image), spatial versus temporal resolution, geometric distortion, spectral sensitivity, shading and linearity. A discussion of these considerations is beyond the scope of this monograph, but an excellent source for this information is the book *Video Microscopy: The Fundamentals* by Drs Shinya Inoué and Kenneth Spring.

The following calculation demonstrates the need for maximizing light throughput in a microscope and using sensitive detectors. The light flux delivered to a specimen on the stage of a microscope from a Xe lamp, 10 nm band-pass, 490 nm interference filter (75% transmission), reflected by a 90% efficient dichromatic mirror into a 100× objective lens is approximately 1.5×10^2 W cm^{-2}. This is equivalent to a photon flux of 0.36×10^{21} photons cm^{-2} or about 10^3 times the brightness of the sun. The amount of emitted fluorescence will be dependent on the molar absorption cross-section of the fluorophore, its quantum yield, the excitation photon flux and the rate of photodestruction of the fluorophore. Using a single molecule of fluorescein as an example, the fluorescence emission rate (FER) turns out to be 1.08×10^5 photons sec^{-1}, and the number of fluorescent emissions before photodestruction (FEP) equals 3.6×10^4. The time to destruction of the fluorophore (FEP/FER) = 0.33 sec. However, the maximum emission rate is limited by the lifetime of the excited state of the fluorophore (4.4 nsec for fluorescein), further decreasing the time available for destruction of the fluorophore to 0.20 sec.

The efficiency of the detection of emission is dependent on the efficiency of the light collection and detection path. The efficiency of detection (in the ideal situation) of an objective lens is dependent on its transmission and NA. Because fluorescence emission occurs over 360° (in three-dimensional space) and only a small fraction of emitted light is collected by the objective, an objective lens with a NA of 1.4 and an efficiency of light transmission of the objective lens of 95% (0.95) can, with this NA, collect approximately 30% (0.30) of the emitted light. If we assume the efficiency of the dichromatic mirror is 85% (0.85) and the barrier filter is 90% (0.90) efficient, the overall detection efficiency is 0.22. Thus, for emission of 3.6×10^4 photons, 8×10^3 reach the detector. Detector quantum efficiency varies from 0.1 to 0.3, resulting in an overall detection efficiency of 2.2–6.6% or 800–2400 photons in 0.33 sec. In the real world, fluorophore concentrations employed are greater than that of a single fluorophore molecule. Assuming the volume of a cell is 2 picolitres and a dye concentration of 10 µM, the detected photon flux would be 0.8–2.4×10^9 photons sec^{-1}. Even at this higher concentration, all of the dye would still photodestruct in 0.33 sec. Longer examination of the cell would thus require decreasing the incident light flux, further reducing the number of photons detected per second.

A number of very sensitive detectors are available for fluorescence microscopy, and can be divided into photon detectors (photomultiplier tubes) or image detectors (cameras). Photomultipliers are the most sensitive detection devices, capable of single photon detection, but suffer from the fact that they are unable to provide information from anything other than a single spot. In terms of image detectors, ultra-low-light-level video or solid state cameras have been developed to provide highly sensitive detection of fluorescence with rapid temporal and high spatial resolution. A variety of these detectors exist, from single stage silicon intensified targets (SIT), two-stage SITs (ISIT), microchannel plate (MCP) intensifiers, charged coupled device (CCD) detectors (cooled or non-cooled, slow scan) and gated intensifiers, where the voltage to the detector can be turned on and off in the order of nanoseconds.

Each of these imaging detectors differs in terms of its sensitivity, linearity, wavelength sensitivity, and temporal and spatial resolution; selection of an appropriate detector will be usage dependent. SIT cameras operate by effectively multiplying input photons (emitted fluorescence) which impinge upon a photocathode, which then impact upon a silicon target where they are read out by a scanning electron beam in a vidicon tube. SIT cameras are useful for detection of moderately low levels of light, but cannot see what cannot be seen by the dark-adapted eye. ISITs contain a SIT fibre-optically coupled to an image intensifier, and have the theoretical capacity to image single photons, although at the expense of low signal-to-noise. This type of camera has found broad application in studies of live cell qualitative imaging where speed is not essential. More recent designs of this class of cameras have relied on the use of MCPs as electron multipliers. Cooled CCD cameras have lower shading and geometric distortion, larger dynamic range, high quantum efficiency, on-chip integration and better linearity, resolution and signal-to-noise than intensified cameras. However, these devices suffer from slow read-out times and a relatively high level of baseline noise. Speed of data acquisition can be increased by 'binning' (combining) picture elements (pixels) on the CCD chip, but at a cost of lower resolution. Intensified CCD cameras [a combination of an image intensifier and a solid state detector (i.e. CCD)] provide a reasonable cost, fast, sensitive detector that doesn't suffer from many of the problems associated with SITs and ISITs. Unlike tube cameras, this combination has essentially no lag (persistence of image information on the video detector or monitor two or more frames after excitation is turned off), but still has a slow read-out time. Position-sensitive detectors exist which can amplify and image single photons, but suffer from extremely low temporal resolution (they require minutes to accumulate a full image), and are rather expensive. A more complete discussion of these cameras can be found in Chapter 6.

2.7.5   *Specimen chambers and stages*

Petri dishes can be used to examine live cells at low and medium magnification on inverted microscopes. Cells can also be fixed on the petri dishes or glass coverslips and examined in the microscope. When it is necessary to follow the physiology of living cells and tissue by direct microscopic observation, sealed environmental chambers are employed. An ideal chamber should be suitable for use with all types of microscope stages, objective lenses and condensers. All parts of the chamber must be biologically inert, made of non-toxic materials (e.g. glass, stainless steel or Teflon) and the chamber design must not be so complex as to constrain practical use. It is imperative to maintain a stable environment in which temperature, pH, CO_2 level and osmolarity are all tightly regulated. There have been numerous reports describing the design of perfusion chambers; most of these chambers have been designed for limited or specialized biological or optical research requirements and may not be applicable to all objectives or microscopes.

Air-curtain incubators (a hair dryer) coupled with air-tight perfusion chambers are most commonly employed for this purpose. However, with this device it is not easy to maintain constant, uniform specimen temperature, particularly when used in conjunction with an immersion objective, as temperature gradients are created by heat absorption through the contact area with the objective. Air-curtain incubators provide a current of warm air that may be turbulent or may be disturbed by room air currents, resulting in variable heat delivery to the specimen. Another type of heat source is an IR light, but care must be taken because the IR radiation can affect low-light-level microscope image detectors. An IR filter is usually sufficient to take care of this problem.

Temperature-controlled perfusion chambers are also available. The perfusion chamber temperature can be maintained by a water channel (from a heated water bath) machined into the chamber. One design takes the bottom of a petri dish and bonds it to a 0.5 mm glass coverslip to allow investigators to culture cells in the petri dish and transfer them directly to the microscope stage. Should a sealed chamber prove unusable, some chambers have been developed which enclose the whole microscope stage and maintain the temperature and CO_2 levels constant.

Perfusion of the chamber at a constant rate with fresh media can be performed with a small syringe infusion pump, preferably made of glass material, to avoid CO_2 leakage. The perfusion characteristics of the chamber are directly related to the flow rate of the media. A flow rate of approximately 1 ml hr^{-1} results in a clean sweeping action of the perfusate through the chamber. Lower flow rates (< 1 m hr^{-1}) result in diffusion of the perfusate in the chamber, and a higher flow rate (> 1 ml hr^{-1}) results in vortical mixing.

Manual or computer-controlled motorized X–Y stages are available and selection depends on the application. Motorized stages are most useful when you wish to observe many samples repeatedly over an extended period of time, particularly when the specimen must be repeatedly removed from the stage (e.g. for restaining or further incubation); once returned to the microscope stage, the computer can then automatically relocate the same cells that were previously imaged. Computer controlled X–Y stages are accurate to within 0.1–1 μm resolution, and their precision-machined mechanical parts provide good protection against wear. Limit switches and high precision, hardened and ground crossroller bearings are required for smooth operation and provide excellent load carrying capability.

3 Practical fluorescence microscopy

3.1 Köhler illumination

The goal of the illumination portion of the microscope is to provide bright and even illumination across the field of view while not illuminating any areas outside of the field of view. This reduces scattered light and photodamage. To accomplish this, the microscope is aligned such that the illumination set of planes are separate from the set of planes that carry the specimen information. The surface of the excitation light source is imaged into the plane of the condenser aperture diaphragm allowing control of the aperture diaphragm, and the field diaphragm is placed in the plane where the back image of the stage object is formed, allowing adjustment of the area of illumination on the stage. This allows the field diaphragm to control the size of the illuminated area. This optical design is named after its inventor, and is known as Köhler illumination (*Figure 3.1*). The advantages of Köhler illumination are that the back focal plane of the objective is fully illuminated, providing a microscope field that is homogeneously bright. Since the specimen is illuminated by a converging set of wavefronts, each arising from separate points of the light source imaged into the condenser aperture, maximal lateral and axial resolution, optimal contrast and reduction of flare from the microscope optics (without any vignetting) are obtainable.

3.2 Arc lamp alignment

To align the arc (e.g. Hg or Xe) lamp, remove the arc lamp housing from the microscope and project the image of the bulb on a wall. Focus the electrodes of the lamp with the collector lens. Once focused, move the mirror so as to remove the mirror completely from the field of view and

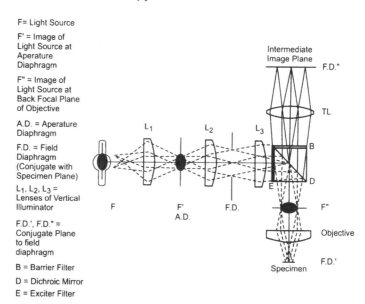

F = Light Source

F' = Image of Light Source at Aperature Diaphragm

F" = Image of Light Source at Back Focal Plane of Objective

A.D. = Aperature Diaphragm

F.D. = Field Diaphragm (Conjugate with Specimen Plane)

L_1, L_2, L_3 = Lenses of Vertical Illuminator

F.D.', F.D." = Conjugate Plane to field diaphragm

B = Barrier Filter

D = Dichroic Mirror

E = Exciter Filter

Figure 3.1. Köhler illumination in a reflected light infinity corrected objective system. A collecting lens (L_1) is placed in front of the light source to make sure that the image of the centrable aperture diaphragm (F' A.D.) is focused on the back aperture of the objective (F"), which is then focused on the specimen. Adapted with permission of Olympus-America, New York, from Abramowitz, M. (1993) *Fluorescence Microscopy: The Essentials.*

centre the true image of the arc such that the electrodes are aligned both vertically and horizontally (*Figure 3.2a*). Next, displace the arc image horizontally and adjust the mirror until the mirror image is beside the arc image and almost touching it (*Figure 3.2b*). Reattach the lamp housing to the microscope and focus the lamp at the plane of the objective aperture diaphragm.

As previously mentioned, these lamps produce high intensity light, exist under high pressure and get very hot during use. Therefore, the following items should be kept in mind when using these light sources. **Never:** (a) use the lamp outside of its housing; (b) look directly at the lamp; or (c) expose skin to the lamp (it could burn and/or cause skin cancer). Lastly, change lamp bulbs only when the lamp is cold (it might explode when hot).

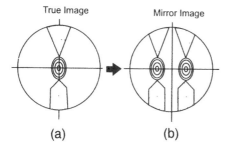

(a)   (b)

Figure 3.2. True and mirror image of lamp bulb as they should appear during lamp alignment. Adapted with permission from Spring, K., Marine Biological Laboratories. Optical Microscopy and Imaging in the Biomedical Sciences course notes.

3.3   Optimizing image brightness

The real advantage of fluorescence microscopy is its ability to detect fluorescent objects because a fluorescent object is very bright relative to the background (which is dark). Optimizing image brightness involves maximizing the principles already discussed in Chapter 2. To summarize, it is critically important to be able to supply the sample with sufficient light energy for excitation at the appropriate fluorophore-dependent wavelength, and to be able to observe the emitted fluorescence without any contamination from the excitation light. The light source must provide a large amount of excitation energy in very narrow ranges of the spectrum; for this purpose, so-called line emitters, usually high-pressure Hg lamps, are used. Appropriate filters which transmit the required wavelengths but block the unrequired ones should be employed. Given that the energy of the excitation is far greater than the energy of the emission, this can be difficult to accomplish, since the intensity of the excitation light is many times higher than that of the emission light. The objectives should have high transmission from the UV through to a large part of the visible spectrum. As we know (*Figure 2.1*), fluorescence emitted from the specimen is radiated in all directions. Objectives with the highest NA should be employed. A good rule of thumb is that if the objective aperture is doubled in size, approximately four times more fluorescence light can be gathered. Oil immersion also eliminates the loss of light caused by light reflection on the surfaces. Lastly, none of the optical components of the microscope should be autofluorescent. Autofluorescence will increase the brightness of the background and reduce the contrast in the image.

3.4   Optimizing image resolution

- Make sure that the objective lenses are clean (see below). A fingerprint on the front lens of an air objective can scatter light and lower contrast. Oil-immersion lenses soiled with residues of resin or emulsions (e.g. oil and water) will also be limited in their ability to achieve optimal resolution.
- It is very important that coverslips used with high NA dry objectives (without immersion oil) have the standard thickness of 0.17 mm; if a different thickness than the standard one is used, the quality of the optical image will degenerate more as the NA of the objective increases. A good rule of thumb is with objectives with NA greater than 0.7, the thickness of the coverslip can vary by ± 0.01 mm (from 0.17 mm) and still provide a high resolution image. When the NA of

the objective is between 0.3 and 0.7, coverslip thickness can vary by ± 0.03 mm and still provide high resolution images.

- Use only oil with the proper refractive index (η = 1.51). Make sure the oil is PCB-free and has very low level autofluorescence. The image will be markedly impaired if air bubbles are contained in the immersion layer. To avoid these bubbles, it is vital to apply the oil in the right way.
- To reduce glare (unwanted reflected light) and improve contrast, close down the field diaphragm.

3.5 Photobleaching

An important quenching phenomenon in the broadest sense is that of photobleaching. Since photobleaching or 'fading' causes loss of fluorescence signal, it is a critically important factor for fluorescence microscopy, yet is a poorly understood phenomenon. As a point of reference, the quantum yield for photobleaching fluorescein at subsaturating intensities is about 3×10^{-5}. This means that, on average, a fluorescein molecule will emit 30 000–40 000 photons during its 'photochemical lifetime'. This is true no matter how the excitation is delivered – in pulses or continuously. Therefore, at low excitation energy levels, photobleaching is not prevented, but the rate of photobleaching is merely reduced.

An important class of bleaching events are photodynamic, that is, they involve the interaction of the fluorophore with light and oxygen. For example, a fluorophore once residing in its lowest excited energy state could undergo intersystem crossing to the long-lived triplet state and in that state be oxidized by proximate molecular oxygen (*Figure 1.2*). The triplet state is most important for photochemistry because its long lifetime allows for more reactions with components of the fluorophore's environment. Alternatively, the excited triplet state could 'sensitize' singlet oxygen production by collision with molecular oxygen, energy transfer and concomitant return of the fluorophore to the ground state. Singlet oxygen is highly reactive and can react with the fluorophore to bleach it. The amount of photodamage will depend on the concentration of molecular oxygen and the distance between the fluorophore, molecular oxygen and any other cellular constituents. The distance over which singlet oxygen can induce damage can be calculated based upon the rate of diffusion in three dimensions $(6Dt)^{1/2}$. Assuming that the singlet oxygen lifetime, t, is about 1 μsec and its diffusion coefficient, D, is 10^{-5} cm^2 sec^{-1}, singlet oxygen could cause photodamage to neighbouring structures provided they were within about 700 Å of the dye.

From a theoretical standpoint, whether the excitation energy is delivered in a pulsed or continuous fashion should not affect the photochemical lifetime. However, empirical evidence suggests that short bursts of high intensity provide an advantage in reducing photobleaching. The mechanisms underlying these empirical observations are not clear, but it may be that very brief, sub-microsecond pulses could allow interaction with only the 'nearest neighbour' oxygen, thereby limiting photobleaching. Conversely, if the excitation flux is too high (as in a laser beam above the milliwatt range focused with a high aperture objective), multiphoton absorption may occur with possible bond photolysis, leading to a photobleached product. However, since it is only possible to get a certain number of photons from each fluorophore during its photochemical lifetime, regardless of how we deliver the excitation, to a first approximation, one or the other approach may have experimental advantages in certain situations. For example, it may be an advantage to get several high signal-to-noise (S/N) digital video images prior to photobleaching using intense short duration pulses. Alternatively, for certain cell physiological events, one may need to illuminate for much longer times to study the time evolution of a phenomenon but at considerably lower S/N in the individual images.

Protection against photobleaching involves reducing exposure time or excitation energy. However, this will also reduce the measurable signal. Solutions of molecules or cell suspensions could also be deoxygenated, although this is not feasible for living cells and tissues. Antifade reagents such as *n*-propyl-gallate or others which are commercially available, can be added to fixed specimens. Singlet oxygen quenchers such as histidine, diphenylisobenzofuran, or crocetin, a water-soluble carotenoid, can also be employed.

3.6   Cleaning the microscope optics

Always work in as clean an environment as possible. Dust is to be avoided at all costs. Keep the microscope covered at all times when not in use. Store all microscope components in appropriate containers and avoid salt water, corrosive solvents and all solvents if possible. Clean the microscope thoroughly after each use. Do not smoke. Before cleaning an objective lens, one should directly inspect the lens using an inverted ocular from the microscope as a magnifier in room light. It is also possible to inspect the objective lens by placing the lens under a dissecting microscope and focusing on the lens to observe any imperfections in the lens.

The first rule of cleaning microscope optics is not to touch the lens surface if at all possible, not even with lens paper. Dust particles in the air and even components of cleaning tissues can contain abrasive materials that will scratch the lens surface. The best way to clean an objective lens is to take a previously unused Q-tip and soak it in solvent. A mixture of 55% acetone/30% ether/15% ethanol, pure alcohol, the glass cleaner Sparkle, distilled water, lighter fluid or a well shaken 1:1:1 mixture of chloroform, alcohol and water can be used. Avoid benzene, toluene and xylene. The organic solvents are usually more effective, but bear in mind that some of these solvents are carcinogens, flammable and need to be used in a well ventilated environment with no nearby flames (i.e. no smoking). Very gently roll the solvent-laden Q-tip over the surface of the lens once. Take another fresh Q-tip dipped in solvent and pass it over the lens very gently, barely touching the lens surface. Another option is to place a drop of solvent on to the objective lens (without directly touching the lens), and then gently blot the drop of solvent up using lens paper. Again, do not touch the lens directly.

Other important facts to bear in mind regarding the usage of objective lenses is to only use oil on lenses designed to be used with oil. It is a tedious process to clean immersion oil off a dry lens so as not to damage the lens. Objectives that are not in use should be stored properly in the container they came in. If you are using oil-immersion objectives and scanning a number of slides, it is not necessary to clean the objective between each slide. Rather, wipe off excess oil with lens paper and place the new slide (with a drop of oil on it) on to the objective such that the oil on the slide and the residual oil on the objective meet without producing air bubbles.

3.7   Further tips for optimal fluorescence microscopy

The choice of microscope type (upright vs. inverted) is dependent upon cost, the sample being observed, experimental manipulations to be performed, type of chamber required to hold the specimen, and accessory optics required to perform fluorescence, transmitted or reflectance microscopy. Inverted microscopes are better when live cell or tissue experiments are being undertaken, as they allow easier manipulation (e.g. fluid exchange, micro-injection, electrophysiology) of the cells or tissue. However, to use an inverted microscope for phase or bright-field imaging, a long working distance condenser is required. Another important concern when choosing an inverted microscope is to ensure that the focusing mechanism does not move the dichromatic mirrors relative to the excitation light path, as this can result in misalignment of the excitation beam. A solid stage that contains mounts to hold both glass slides and environmental chambers, or which can easily be adapted for this purpose, is advantageous for an inverted microscope.

Fluorescence is best viewed in a dark environment after allowing your eyes to adapt to the dark. When the specimen is not being viewed or photographed, block out the excitation light using a filter slider or shutter in the fluorescence illuminator in order to avoid undesired bleaching by the excitation light. Make sure the immersion oil you are using is free of autofluorescence. In addition to immersion oil, autofluorescence can come from the cells, tissues or biological fluids being examined. Using fluorophores with excitation more than 500 nm will diminish autofluorescence, and because light scattering by cells and tissues is reduced at higher wavelengths, greater penetration of the excitation light into the specimen can be achieved. Fluorescence filters are sensitive to the heat emitted by the illuminator. Never remove the integrated heat protection filter unless your know that it will not transmit the wavelengths of light you are using for excitation. The illuminator must be readjusted occasionally – in all cases after lamp replacement. Lastly, unbound fluorochromes should be removed from the specimen, for example by washing them out. The 'contrast' in the fluorescence image is generated by the dark background only, which in turn is unnecessarily brightened by excess fluorochromes.

4 Applications of fluorescence microscopy

4.1 Varieties of fluorescence microscopy

Over the past 40 years, fluorescence microscopy has undergone a rapid and substantial evolution in its design and in the number of applications for which it can be used. *Table 4.1* lists the various types and applications that currently exist for fluorescence microscopy. The more useful of these will be discussed briefly in terms of their optical design and application; digitized video microscopy will be discussed in Chapter 6.

4.1.1 Immunofluorescence

Immunofluorescence is used to visualize specific cell or tissue components – it combines the specificity of antibodies with the sensitivity of fluorescence. It is based on the fact that it is possible to produce highly specific antibodies against individual cell constituents (antigens), to label these antibodies with fluorescent molecules and then visualize these fluorescent antibodies after they have been applied to microscopic specimens. The antigen–antibody reaction is highly specific in that an antibody will only react with the antigen which stimulated its production.

Antibodies can be covalently labelled with fluorescent molecules without completely losing their immunological specificity (*Figure 4.1*). Antibodies labelled with fluorescent molecules are known as conjugates, and can be applied to either cellular or tissue samples whereupon, if the particular antigen is present, the conjugate will bind to the antigen allowing visualization of the antibody (and hence antigen) distribution. The degree of labelling of an antibody molecule with fluorescent molecules is characterized as the ratio of the number of fluorochrome molecules to the number of protein (i.e. antibody) molecules (F/P ratio). A high F/P ratio will result in a brighter fluorescence image, but one with higher non-specific fluorescence. Also, as the F/P ratio increases, immunological specificity decreases. An F/P ratio of 1–2 is preferred.

Table 4.1. Types and applications of wide-field fluorescence microscopes

Types	Applications
Immunofluorescence	Identification, localization and visualization of cell and tissue components
Fluorescence *in situ* hybridization (FISH)	Quantitative detection and localization of specific DNA or RNA sequences
Fluorescence/differential interference contrast	Dynamic observation of fluorescent cellular components in conjunction with cellular morphology
Automated fluorescence image cytometry	Automatic identification and quantification of fluorescent material in cells and tissues
Fluorescence recovery after photobleaching (FRAP)	Lateral diffusional mobility of cellular components
Total internal reflectance fluorescence microscopy (TIRF)	Visualization of fluorescent objects within 200 nm of substrate
Fluorescence resonance energy transfer microscopy (FRETM)	Interaction of cellular components at the molecular level
Digitized fluorescence polarization microscopy (DFPM)	Rotational mobility of cellular components
Fluorescence lifetime imaging microscopy (FLIM)	Two- or three-dimensional imaging of fluorescent lifetimes in cell and tissue
Intravital fluorescent microscopy	Observation of fluorescent markers of organ physiology *in vivo*
Fourier spectroscopy and spectral dispersion microscopy	Complete spectral analysis from microscopic samples
Delayed luminescence microscopy	Reduce sample autofluorescence; clinical diagnostics
Green fluorescent protein	*In situ* visualization of cell and tissue component distribution and interactions
Digitized video microscopy	Measurement and manipulation of real-time intact cell and tissue physiology; observation of multiple cellular activities in same specimen
Ratio imaging	
Multiparameter	
Caged compound/photoactivation	

Two approaches are commonly used in immunofluorescence microscopy. Labelling an antibody with a fluorescent molecule and reacting that conjugate with a sample containing the appropriate antigen is known as direct immunofluorescence. While the specificity of this approach can be high (because one antigen can only bind from one to a few antibodies), the overall fluorescence signal can be weak. However, it is possible to enhance the fluorescence signal by increasing the number of antigenic sites available to bind fluorescently labelled antibody. This approach is called indirect (secondary) immunofluorescence (*Figure 4.1*). In indirect immunofluorescence, the antigen is first incubated with an appropriate antigen-specific unlabelled antibody. Then another antibody, which is labelled with the fluorescent molecule and which was prepared using the first antibody as an antigen, is incubated with the unlabelled antigen first antibody complex. Since each antigen is able to bind to multiple antibodies, this approach provides more binding sites for secondary antibody binding and amplification of the signal compared to direct immunofluorescence.

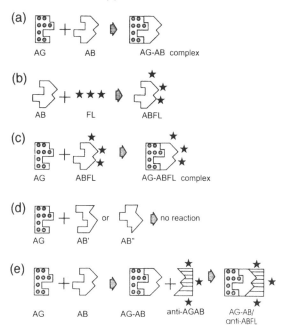

Figure 4.1. Main reactions involved in immunofluorescence. (a) Normal antigen–antibody reaction; (b) labelling antibody with fluorophore (FL); (c) direct immunofluorescence – reaction of fluorophore-labelled antibody with antigen allowing detection of antigen presence or absence and distribution in sample after excitation; (d) no interaction of antigen and antibody because antibody not specific for antigen; (e) indirect immunofluorescence – a second antibody raised against the primary antigen-specific antibody is labelled with fluorophore providing 4–10× amplification of signal.

While the antigen–antibody interaction is highly specific, it is not perfect. Thus, one of the primary problems of immunofluorescence is non-specific fluorescence. The source of this non-specific fluorescence can be primarily from the specimen itself (autofluorescence) as well as secondary fluorescence where interaction of exogenous substances with cellular constituents forms a fluorescing species. Secondary autofluorescence can be due to fixatives (e.g. glutaraldehyde) which form aromatic fluorescing compounds when reacting with cells, while primary fluorescence is usually due to cellular components that fluoresce blue or green when excited by UV light (e.g. tryptophan, flavines, lipofuschin etc.). There are a number of steps that can be taken to try to minimize primary and secondary fluorescence. Use of wavelength selection devices (excitation and emission) which are highly specific in their transmittance for the particular fluorophores being employed will decrease non-specific fluorescence, although at the price of a weaker intensity image. Use of non-fluorescent fixatives (e.g. methanol, ethanol, acetone, formaldehyde, paraformaldehyde) and extensive washing between each step of the immunofluorescence process will help limit autofluorescence. Non-specific reactions of the fluorescent conjugate with non-antigen components of the specimen can also lead to secondary fluorescence. Non-specific fluorescence can be reduced by removing all of the non-conjugated (free) fluorescent molecules from the conjugate; this is usually done by gel filtration or ion-exchange chromatography. Proper storage of the labelled conjugate (aliquoted and stored at −20°C or freeze dried) is also required; repeated freeze–thawing disrupts the covalent bonds between the fluorescent molecule and protein. The use of mounting medium which contains agents that prevent photobleaching (see Chapter 2) or enhance fluorescence [i.e. glycerol–phosphate-buffered saline, pH 7.2, 9:1 (v/v)] can also decrease non-specific fluorescence.

4.1.2   *Fluorescence* in situ *hybridization (FISH)*

FISH is very similar to immunofluorescence except that FISH allows the direct visualization and localization of DNA and RNA sequences on chromosomes, in cells and in tissue. FISH is based on the hybridization between target sequences of single-stranded DNA of chromosomes or nuclei with fluorescently labelled complementary sequences (cDNA). The advantages of FISH over other types of *in situ* hybridization approaches are that it is a fast, reliable, sensitive and quantitative method, which can detect as little as one copy of a gene in an individual cell.

As is the case with immunofluorescence, both direct and indirect FISH are possible. In direct FISH, fluorescent molecules (e.g. fluorescein) are directly conjugated to the nucleic acid probe, allowing probe–target hybrids to be visualized through the microscope. Success of this approach requires the use of fluorophores of high quantum yield, the ability of the probe–fluorophore linkage to survive the hybridization process and that the conjugated fluorophore does not interfere with the hybridization reaction. Fluorophores currently used in FISH include AMCA (amino-methyl-coumarin-acetic acid, ex. 399, em. 446); fluorescein (ex. 495, em 515); CY3 (cyanine 3, ex. 552, em. 565); rhodamine (ex. 554, em. 580); and Texas red (ex. 590, em. 615). Using a combination of these fluorophores, triple label FISH is possible. It is also possible to increase the number of detectable targets by using combinatorial labelling, where the number of detectable targets is 2^n-1, where n is the number of colours. The ratioing of fluorescence intensities of various fluorophore combinations (combinatorial labelling) allows the simultaneous identification of up to 12 different DNA sequences if individual targets are spatially separated (e.g. metaphase chromosomes).

Indirect FISH is performed by labelling the probe with a molecule to which many other molecules can bind (*Figure 4.2*). For example, if antibodies are available against the fluorophore which is directly conjugated to the nucleic acid probe, one can amplify the detectable signal and change direct FISH into indirect FISH. There have been a number of compounds that have been identified that can be used in indirect FISH, but the two most popular are biotin (detected with strepavidin) and digoxigenin (detected with anti-digoxigenin antibodies). The digoxigenin labelling procedure is based on the use of digoxigenin, which is isolated from digitalis plants and is linked to the C-5 position of uridine nucleotides via an 11 carbon atom spacer. These digoxigenin labelled uridine nucleotides can then be incorporated into nucleic acid probes (at a known density, usually every 20th to 25th nucleotide), and once hybridized to the DNA of interest, detected with anti-digoxigenin antibodies conjugated to fluorescent molecules. Biotin can also be conjugated to uridine and detected with fluorescent strepavidin, which has a very high affinity for biotin.

There exist a number of procedures to label DNA probes. These include random priming, where end-labelled small oligonucleotides are hybridized to denatured double-stranded DNA which is used as a template. Synthesis of the new strand is performed by using Klenow polymerase in the presence of unlabelled or labelled deoxynucleotide triphosphates (dNTPs). This approach works best for large DNA sequences and results in probes that vary between 200 and 1000 bases long. Nick translation is another approach for labelling DNA. Nicks are introduced in double-stranded DNA followed by addition of the enzyme DNA polymerase (holoenzyme) and labelled dNTPs. Polymerase chain reaction (PCR) can also be used to generate labelled cDNA by using either end-labelled oligonucleotide primers or labelled dNTPs in the PCR reaction. It is also possible to attach probes to the 3′ end of the oligonucleotides using fluorophore-, digoxigenin- or biotin-conjugated uridine nucleotides. A variant of this procedure is to attach a string of derivatized uridine molecules to the tail of oligonucleotides (so-called tail labelling; *Figure 4.2*). This allows a large number of labelled nucleotides to be added to a small oligonucleotide probe, usually resulting in increased sensitivity, specificity and enhanced signal.

4.1.3 *Fluorescence/differential interference contrast (DIC)*

The ability to image living cell shape and size and quantitate the cellular distribution of fluorescent material at the same time represents a powerful approach to understanding the role of specific cellular constituents in overall cell function. In DIC microscopy, contrast is generated due to a very small shear or splitting of an incoming polarized beam of light by the specimen. Regions of a microscope specimen that have rapid changes in thickness or refractive index generate the most contrast (i.e. at the edges of objects). The resulting shadowed image generates a three-dimensional appearance of the specimen. DIC requires the use of polarizing optics including a polarizing element called an analyser in the light path between the objective lens and detector, which typically has very low transmittance and thus results in unacceptable light loss when examining fluorescent objects. Fortunately, there is a solution (*Figure 4.3*). The requirement for an analyser in the detection light path

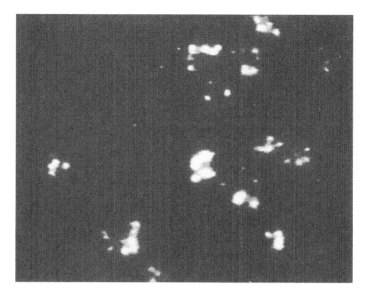

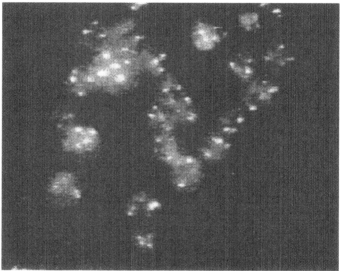

Figure 4.2. Fluorescence *in situ* hybridization (FISH). Caski cells which express 500 copies of human papillomavirus 16 (HPV 16) were fixed and processed for the detection of HPV 16 DNA using FISH. Nuclei were counterstained with Hoescht 33258 dye to visualize nuclear morphology. HPV 16 is integrated into multiple sites in Caski cell DNA and thus many dots of staining can be seen in each nucleus. Top: random primed digoxigenin HPV 16 cDNA visualized with fast red alkaline phosphatase anti-digoxigenin antibody. Bottom: FITC tail-labelled HPV 16 oligomer.

can be met by the dichromatic mirror. In fluorescent microscopes, dichromatic mirrors are oriented at 45° relative to the incident beam and thus can act as a polarizer, resulting in the transmission of two different wavelengths of light in planes perpendicular to one another. This allows both transmitted and fluorescence observation of the same sam-

(a)

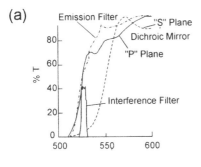

(b)

 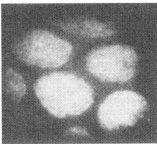

Figure 4.3. Simultaneous DIC and fluorescence microscopy. (a) The dichromatic mirror acts as a polarizer, resulting in the transmission of two different wavelengths of light in planes perpendicular to one another. This wavelength-dependent polarization effect only operates over a very narrow range of wavelengths and it is important to evaluate carefully the spectral separation between the two different polarized transmission planes and the steepness of the cut-off of the dichromatic filter. (b) The left hand picture shows a DIC image of *Necturus* gallbladder epithelium; the right-hand picture is an image of acridine orange stained nuclei. Adapted with permission of Wiley-Liss, New York, from Herman B. and Jacobson, K. (1990) *Optical Microscopy for Biology.*

ple, and mitigates the low transmittance obtained when polarization optics are required for DIC. This wavelength-dependent polarization effect only operates over a very narrow range of wavelengths and it is important to evaluate carefully the spectral separation between the two different polarized transmission planes and the steepness of the cut-off of the dichromatic filter.

4.1.4   Automated fluorescence image cytometry (AFIC)

The use of fluorescent stains coupled with computerized vision microscopy systems enables the screening of stained specimens with reliable and accurate detection of almost all cells, automatically. In addition, the use of fluorescent stains attached to antibodies or to nucleic acid probes provides highly specific and sensitive probes to detect disease markers in clinical specimens. The use of such fluorescent disease-specific markers results in only the marker in the specimen being excited and appearing as signal in the image, resulting in images with high signal-to-noise (S/N) ratios which serve to increase the sensitivity of detection.

A typical AFIC consists of an epi-fluorescence microscope, low-light-level camera and computer. The images from the microscope are captured by the camera, digitized and stored in computer memory. For precise quantification, the images are corrected for background (by subtracting an image not containing stained objects) and variations in the efficiency of the system over the imaging area (shading). Shading is assessed by measuring a standard fluorescent object at different positions over the imaging area. Cell nuclei are automatically detected by recording images of them, correcting the images for background and shading and segmenting the images into regions corresponding to stained nuclei and (unstained) background. In addition, the AFIC has a computer-controlled stage, which automatically records the location of the images relative to a fixed reference point to a precision of $1\mu m$. This feature allows the user to remove the slide from the stage (for further staining etc.) and then re-image the same scenes as before. Also, the focus drive of the microscope is under computer control in order to maintain optimal focus.

The ability of AFIC to automatically scan and identify fluorescent objects on a slide is based on the fact that, when stained with fluorescent dyes, the objects are very bright compared to the background, allowing highly reliable identification of objects on a slide using image segmentation techniques in addition to object intensity, shape and size. Processing of the microscope images usually includes filtering, object identification using adaptive, gradient-weighted thresholding methods, edge detection and various other properties (e.g. location in the image, area, perimeter, total fluorescence intensity) of the objects in the image (*Figure 4.4*).

4.1.5 Fluorescence recovery after photobleaching (FRAP)

This technique allows the translational mobility (lateral diffusion) of fluorescently labelled molecules to be monitored in very small (2–5 μm) regions of a single cell's surface or cytomatrix. The basic concept of a spot photobleaching measurement is as follows (*Figure 4.5*). The emission (F_i) produced by laser excitation of a fluorescent label on or in a region of a membrane, surface or thin volume is measured. Next, an intense, short photobleaching pulse of light is directed to the same region of the specimen, rapidly destroying a substantial amount of the fluorescence in that region. After this photobleaching pulse, the end of which is defined as time zero, the recovery of fluorescence within the photobleached region is measured as a function of the time with a beam coincident with the photobleaching beam but of the same intensity as that used to measure F_i. This fluorescence intensity during the recovery phase ranges from F_0

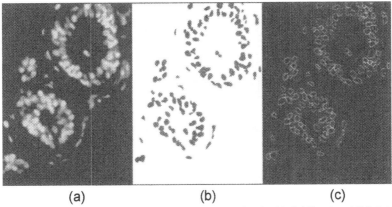

(a)   (b)   (c)

Figure 4.4. Analysis of an image of breast tissue stained with DAPI using AFIC. (a) The original 1280 × 1024 pixel image. (b) Result from full analysis of the third of the image containing most of the clustered nuclei. (c) The edges of the detected regions. Adapted with permission of Wiley-Liss, New York, from Lockett, S. and Herman, B. (1994) Automated detection of clustered, fluorescent-stained nuclei by digital image-based cytometry. *Cytometry* **17**: 10.

(at time zero) to F_∞, the time when the recovery is judged complete. The kinetics of recovery can be empirically characterized by the time required to reach 50% of complete recovery ($\tau_{1/2}$). The mobile fraction [mf $= (F_\infty - F_0)/(F_i - F_0)$] characterizes the extent to which the initial fluorescence, F_i, is regained. Fluorophores immobile on the time-scale of the measurement are indicated by an mf of less than 100%. If spontaneous recovery of fluorescence does not occur and the probe beam does not itself induce photobleaching, the recovery kinetics can be related to the rate of lateral transport of neighbouring unbleached fluorophores into the previously bleached region in a straightforward manner. For example, in the case of diffusion-limited fluorescence recovery in two dimensions, the diffusion coefficient is directly proportional to the square of the laser spot radius in the specimen plane and inversely proportional to the half-time for recovery, $\tau_{1/2}$.

In a typical FRAP experiment, a photomultiplier measures the total brightness levels of the bleached region in the sample. In video-FRAP, the photomultiplier tube is replaced by a low-light-level detector so that the FRAP data is acquired just as if the investigator were viewing the experiment through the microscope. Thus, video-FRAP offers more information in the form of a two-dimensional array of brightness, allowing analysis of such data for evidence of anisotropic diffusion and flow.

4.1.6 *Total internal reflectance fluorescence microscopy (TIRF)*

When a light beam moves through two different media which have different refractive indices, the Fresnel formula states that a light beam propagating in a medium with a higher refractive index will be totally

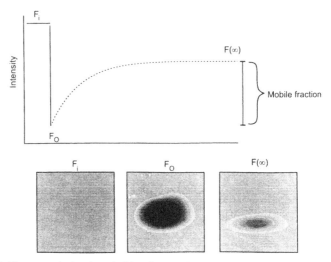

Figure 4.5. Photometric (top) and video (bottom) measurement of fluorescence recovery after photobleaching (FRAP). A small region of the sample containing fluorophore (F_i) is irreversibly bleached by a strong laser beam (F_0) and a much weaker laser beam then monitors the recovery of fluorescence into the bleached area as a function of time. The mobile fraction is defined as the difference between F_∞ and F_0.

reflected at the boundary between the medium with a lower refractive index if the angle of incidence equals or exceeds a critical angle (e.g. for quartz/water: 65.4°). However, a so-called evanescent wave enters the second medium to a depth of roughly 0.1 μm and is capable of exciting only fluorophores which are located at the interface or within this small depth (*Figure 4.6*). This special type of illumination has been used for the measurement of adsorption–desorption kinetics and surface diffusion rates of proteins at the quartz–aqueous buffer interface. It has also been used to study molecules on the ventral (substrate attached) membrane of tissue culture cells.

4.1.7   *Fluorescence resonance energy transfer microscopy (FRETM)*

By combining optical microscopy with FRET, it is possible to obtain quantitative temporal and spatial information about the binding and interaction of proteins, lipids, enzymes, DNA and RNA *in vivo*. Because energy transfer occurs over distances of 10–100 Å, a FRET signal corresponding to a particular field (or pixels) within a microscope image provides additional information beyond the limit of resolution of the microscope down to the molecular scale. FRET microscopy is particularly useful in examining *temporal* and *spatial* changes in the distribution of fluorescently conjugated biological molecules in living cells.

In practice, there are two ways to perform FRET microscopy, through the use of steady state or time-resolved methods. In steady state FRET imaging, an increase in the proximity of the donor and acceptor fluo-

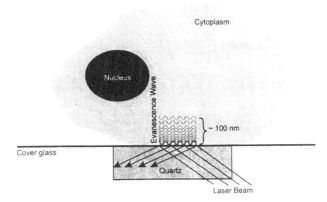

Figure 4.6. Total internal reflectance fluorescence (TIRF) microscopy. Schematic drawing of the evanescent wave produced by a totally internally reflected laser beam at a quartz–buffer interface. When the 488 nm line of an Ar ion laser is incident at an angle of 72° on a quartz–buffer interface, the depth of penetration of the evanescent wave is approx. 97 nm.

rophores results in non-radiative transfer of the donor excitation energy to the acceptor molecules which results in a decrease of the donor emission, I_D, and an increase in the intensity of emission of the acceptor, I_A (sensitized fluorescence; *Figure 4.7*). By ratio imaging of I_A/I_D (sensitized ratio FRET imaging), it is possible to obtain a two-dimensional FRET image. The value of this ratio depends on the average distance between donor–acceptor pairs. FRET imaging can be performed using a standard fluorescence microscope equipped with excitation/emission filters and sensitive video cameras which are currently used for the popular technique of ratio imaging. It is also possible to measure FRET by measuring the fluorescent lifetime of the donor in the presence and absence of the acceptor. The advantage of time-resolved FRET imaging is that the donor–acceptor distance can be mapped in a more accurate and quantitative manner.

Successful undertaking of FRET microscopy requires that several points be considered.

(i) The concentration of the donor and acceptor fluorophore need to be tightly controlled. Statistically, the highest probability of FRET occurs when a single donor is surrounded by a number of acceptor molecules. In order to achieve high S/N, high fluorescence signals are preferred. However, too high a concentration of dye can cause self-quenching and disordered biological function.

(ii) Photobleaching needs to be prevented. Photobleaching can alter the donor–acceptor ratio and therefore the value of FRET.

(iii) Ideally, the donor emission spectrum should substantially overlap the absorption spectra of the acceptor.

(iv) There should be relatively little direct excitation of the acceptor at the excitation maxima of the donor. One limitation often encountered in microscopic measurements of FRET using steady state

Resonance Energy Transfer Microscopy

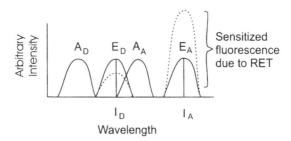

Figure 4.7. Sensitized fluorescence. Resonance energy transfer (RET) results in a decrease (quenching) of the donor emission (dotted line of E_D) when measured at I_D, the wavelength of the donor emission maximum, and a concomitant increase (sensitization) of the acceptor emission (dotted line of E_A) when measured at I_A, the maximum wavelength of the acceptor emission. The ratio of I_A/I_D can be taken as a measure of RET. Adapted with permission of Academic Press, San Diego, from Taylor, D.L. and Wang, Y.-L. (1989) Fluorescence microscopy of living cells in culture. Part B. *Methods Cell Biol.* v. 30.

methods, is that some direct donor emission is detected with the acceptor and FRET filters. Similarly, direct acceptor emission can be detected with the donor and FRET filters.

(v) Both the emission of the donor and acceptor occur in a wavelength range in which the detector has maximum sensitivity.

(vi) There should be little if any overlap of the donor absorption and emission spectrum, thus minimizing donor–donor self-transfer.

(vii) The emission of the donor should ideally result from several overlapping transitions and thus exhibit low polarization. This will minimize uncertainties associated with the K^2 factor.

(viii) The donor must be fluorescent and of sufficiently long lifetime.

(ix) When antibodies are used for FRET imaging, it is necessary to determine whether the antibody reagent itself may be affecting the FRET measurements. For example, because antibody labelling may affect the original molecule's structure and therefore distance information, the correct selection of labelling procedure and antibody needs to be seriously evaluated. Quantitative analysis of FRET data is dependent upon the availability of correct controls, and on the ability to choose the best method available for FRET microscopy measurements.

FRET microscopy has been used to probe cell–substrate interactions, growth factor receptor dimerization, the interaction of electron transport chain components, fusion of endosomes and interaction of viral and cellular proteins. Studies of epidermal growth factor (EGF) receptors have been carried out using time-resolved FRET imaging. EGF receptor clustering during signal transduction was monitored and a stereochemical model for the tyrosine kinase activity of the EGF receptor has been investigated. Time-resolved FRET imaging has also been applied to study the extent of membrane fusion of individual endosomes in single cells. Using time-domain fluorescence lifetime imaging microscopy (FLIM) and FRET, the extent of fusion and the number of fused and unfused endosomes were clearly visualized and quantitated. FRET imaging has also been undertaken to examine the interaction of human papillomavirus (HPV) 16 E6 protein and the tumour suppressor p53 protein.

4.1.8   *Digitized fluorescence polarization microscopy*

Fluorophores preferentially absorb photons whose electric vectors are aligned parallel to the absorption transition dipole moments of the fluorophore (see Chapter 1). Using polarized light, it is possible to selectively excite those fluorophore molecules whose absorption transition dipoles are parallel to the electric vector of the excitation. The transition dipole moments for absorption and emission have a fixed orientation within each fluorophore, and the angle between those dipole moments determines the maximum (limiting) measured anisotropy. Rotational motion (that is, further displacement of the emission dipole from its starting position) which occurs during the lifetime of the excited state, will further lower the observed emission anisotropy. Polarization measurements can be made by placing a polarizer in the excitation light path and an analyser in the emission light path oriented either parallel or perpendicular to the plane of the excitation polarizer.

Microscopic fluorescence polarization measurements on single cells present certain problems not associated with fluorometric measurements. The major difficulties include the depolarizing effects of the microscope optics, the orientational dependence of the observed fluorescence polarization, and the lower fluorescence signals obtainable from individual cells. The depolarization effects of the microscope optics arise from the necessity of using objective lenses with high apertures for maximal collection of fluorescence, rather than a narrow slit of effectively zero aperture. The orientational dependence of the observed fluorescence polarization is due to the fact that the surface of all cells is three-dimensional, rather than a two-dimensional bilayer. Thus, surface curvature, in addition to the rotational mobility of the fluorophore during its excited state lifetime, will determine the final observed depolarization. Correction for the effects of surface curvature requires knowledge of the probe's excited state orientation distribution in order to determine orientation-dependent fluorescence polarization.

4.1.9   *Fluorescence lifetime imaging microscopy (FLIM)*

FLIM allows the sensitivity of fluorescence lifetime to environmental parameters to be monitored in a spatial manner in single living cells. As such, FLIM provides a sensitive technique capable of obtaining data regarding the dynamics and heterogeneous nature of cellular components which cannot be obtained by using the methods mentioned above which are based on time averaging of the fluorescent signal.

To make time- and spatially resolved lifetime measurements at the same time, a time-gated multichannel plate (MCP) image intensifier linked up to a charge coupled device (CCD) camera is used (*Figure 4.8*). The gated MCP is turned on for a very brief interval (e.g. 2 nsec) at some time interval t_1 after the exciting pulse, and the emitted intensity from the gated MCP is acquired on the CCD (which is continually left on). This process is repeated a large number of times (e.g. 10 000) at this time t_1 with the emitted intensity from the gated MCP continually being accumulated on the CCD. The CCD is then read out, the time of the gate window with respect to the excitation pulse is temporally shifted (t_2) and the whole process is repeated. Time-resolved fluorescence images can be directly detected over a short time and stored in a computer.

Recently, FLIM has been used to monitor the extent of membrane fusion of individual endosomes in single cells. By using fluorescence resonance energy transfer techniques and FLIM, the extent of fusion, and

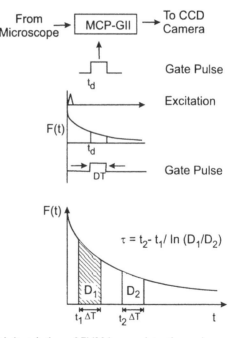

Figure 4.8. Functional description of FLIM image detection using a multichannel plate (MCP) nanosecond gated image intensifier. The fluorescent sample is excited with a picosecond laser pulse and a gate pulse which turns on the voltage to the camera and is delivered at some time ($t_d = t_1$) after the excitation pulse for a certain duration (ΔT). The image is then read off the MCP gated image intensifier on to a CCD camera and the process is repeated except that the time when the gate pulse is delivered to the MCP gated image intensifier is delayed for a longer period ($t_d = t_2$). Adapted with permission of the Society of Photo-Optical Imaging in Biology, Bellingham, from Herman, B., Wang, X.F., Periasamy, A., Kuron, S., Gordon, G. and Wodnicki, P. (1996) *Fluorescence Lifetime Imaging in Cell Biology* **2678**: 88–97.

the number of fused and unfused endosomes were clearly visualized. In another study, FLIM has also been used to monitor levels of cytoplasmic and nuclear Ca^{2+} in cells. FLIM allows quantitative measurements of intracellular Ca^{2+} and pH using non-ratiometric visible wavelength Ca^{2+}/pH-sensitive indicators, because measurement of Ca^{2+} and pH using FLIM is independent of probe concentration and photobleaching. FLIM also allows easy calibration of ion concentrations in cells, something which is not possible with intensity-based measurements. In addition, FLIM also provides the opportunity to study the dynamics of the environment surrounding the ion-sensing probes; two- or three-dimensional images of the fluorescent lifetime of ion-sensitive fluorophores and their rotational mobility (using time-resolved emission anisotropy) can be obtained during response of single living cells to stimulation. This allows the quantitative determination of cellular Ca^{2+} levels, as well as documenting whether any change in probe characteristics (e.g. binding to cytoplasmic or nuclear components) occurs during the measurements.

4.1.10 Fourier spectroscopy/spectral dispersion microscopy

Fourier spectroscopy/spectral dispersion microscopy is a relatively new type of fluorescence microscopy. Spectral imaging allows the user to obtain the spectrum of emitted fluorescence at every picture element (pixel) in a two-dimensional image. Spectral imaging can be used to detect, identify and separate at any location in an image mixtures of several fluorophores simultaneously in one measurement. Spectral imaging (also known as Fourier spectroscopy or interferometric spectroscopy) is based on the use of an interferometer (spectral dispersing element) that separates the emitted fluorescence into two coherent beams while introducing a variable optical path difference (OPD) between the two beams (*Figure 4.9*). The beams are then recombined in such a way as to interfere with each other, and the resulting interference intensity is measured by a detector as a function of the OPD. This provides an intensity versus OPD function which, when Fourier transformed, provides the spectrum. Spectral imaging has very high optical throughput, resulting in high signal-to-noise images with high temporal resolution. To date, spectral imaging has found use in fluorescence microscopy in the spectral identification of multiple fluorophores in a mixture, detection of environmental effects (e.g. pH) on indicator spectra, measurement of naturally fluorescing pigments and FRET by monitoring sensitized emission.

4.1.11 Delayed luminescence microscopy

Advances in fluorescence microscopy and the availability of bioreagents such as monoclonal antibodies and nucleic acid probes have opened up new possibilities for the localization and analysis of proteins and nucle-

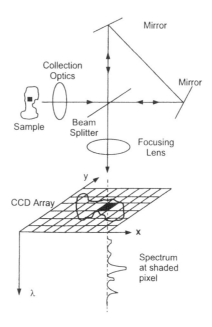

Figure 4.9. Simplified optical diagram of a spectral bio-imaging system based on the Spectra-Cube™ method. The light emitted by the sample is collected by the fore-optics and enters the Sagnac interferometer. The real image (plus interference fringes) is focused on the CCD array, yielding (on Fourier transformation) the spectrum at each pixel. Adapted with permission of Wiley-Liss, New York, from Wang, X.F. and Herman, B. In: Winefordner, J.D., ed. (1996) *Fluorescence Imaging Spectroscopy and Microscopy Chemical Analysis*, v. 157.

ic acid sequences in cells, tissues and chromosomes. Fluorescence assays are increasingly utilized for a variety of applications in both basic and clinical biomedical research, because fluorescent labels are very sensitive, specific, safe (i.e. non-radioactive) and quantitative. In addition, labelling of probes (antibodies or DNA) is relatively easy and yields labelled compounds with high specific activity and good stability. Fluorescence also has the advantage of being able to be used for the simultaneous detection of multiple antigens. However, non-specific auto-fluorescence from solvents, solutes, cell and tissues, fixative-induced fluorescence of cells and tissues and autofluorescence of the optical components of the microscopic system limit the sensitivity of detection. This is a common problem, affecting essentially all studies employing fluorescence microscopy, whether in living cells, fixed tissues or clinical samples, and serves to effectively decrease the signal-to-noise ratio of detection.

Autofluorescence typically has an intensity equivalent to 10^2–10^4 fluorescein molecules; the lifetime of autofluorescence and other non-specific fluorescence is in the range of 1–10 *nsec*. In contrast, phosphorescent and certain other compounds (e.g. lanthanides) exhibit fluorescence having lifetimes in the range of 1 μsec–10 msec. Therefore, if a sample which used phosphorescent or delayed fluorescence compounds as probes, could be excited with a short (nsec) pulse of light, and detection delayed to the micro- or millisecond range, the emitted intensity could be observed without interference from autofluorescence and scattered light, leading to a substantial increase in detection sensitivity (*Figure 4.10*). In time-resolved fluorescence, fluorescence emitted from the probe is measured after a delay during which the background fluorescence (autofluorescence) has completely decayed. The ability to discriminate between specific fluorescence (or phosphorescence) and autofluorescence and other non-specific fluorescence suppresses background by two orders of magnitude, providing sensitivity equivalent to radioimmunoassay.

In practice, a time-resolved fluorescent image of a sample is obtained by applying a sampling gate pulse to the photocathode of the detector after a delay following sample excitation. In the example illustrated in *Figure 4.10b*, following a 30 µsec delay after a 250 µsec duration excitation pulse, a gated camera is turned on for a brief period of 250 µsec. Commercially available long lifetime probes [Y,O,S:Eu, Zn,SiO:MNAs,

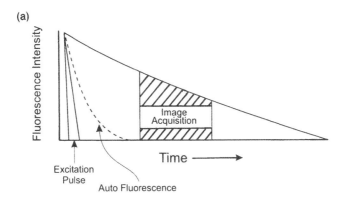

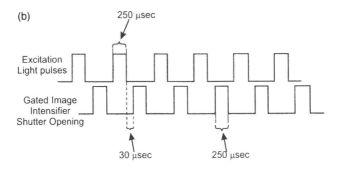

Figure 4.10. (a) Principle of delayed luminescence microscopy. A short pulse of excitation is delivered to a sample containing fluorophores with long lifetimes (i.e. msec). The detection of the emitted fluorescence is delayed for a period of time (approx. 500 µsec) following delivery of the excitation pulse, allowing autofluorescence to decay to baseline. (b) Representation of timing of excitation pulses to the gated image intensifier. Following excitation with laser light of 250 µsec duration, 30 µsec after the excitation pulse the gated image intensifier is turned on for 250 µsec to acquire the time-resolved image. This process is repeated until an image with adequate signal-to-noise ratio is accumulated. Adapted with permission of Preston Gordon Communications, UK, from Periasamy, A., Siadat-Pajouh, M., Wodnicki, P., Wang, X.F. and Herman, B. (1995) *Time Gated Fluorescence Microscopy for Clinical Imaging: Microscopy and Analysis.* pp. 33–35.

ZnS:Ag, the chelate 4,7-bis(chlorosulphophenol)-1,10-phenanthroline-2,9-dicarboxylic acid (BCPDA) completed with lanthanide ions (Eu, Tb, Dy and Sm)] can be used as labels for antibodies or DNA sequences.

4.2   Green fluorescent protein (GFP)

The green fluorescent protein (GFP) of the jellyfish *Aequorea victoria* has been found to fluoresce with fluorescein-like characteristics and is attracting tremendous attention as a strong visible fluorescent reporter molecule which is species-independent and does not require any cofactors or substrates. To date, GFP has been used as a reporter of gene expression, a tracer of cell lineage and as fusion tags to monitor protein localization within living cells.

The GFP chromophore consists of a cyclic tripeptide derived from Ser-Tyr-Gly in the primary protein sequence and is only fluorescent when embedded within the complete GFP protein. Purified GFP is a 27-kDa monomer consisting of 238 amino acids. GFP crystal structure studies demonstrate an 11-stranded β-barrel with a coaxial α-helix containing the chromophore. The chromophore is formed by a cyclization reaction and an oxidation step that requires molecular oxygen. The wild-type (wt) GFP absorbs UV and blue light with a maximum peak of absorbance at 395 nm and a minor peak at 470 nm, and emits green light maximally at 509 nm, with a shoulder at 540 nm. GFP has a quantum yield of about 80% but a lower extinction coefficient than fluorescein. Studies with wt GFP expressed in HeLa cells have shown that the cytoplasmic concentration must be greater than approximately 1.0 mM to obtain a signal that is twice the autofluorescence. This detection limit is equivalent to approximately 4000 molecules per cell in the cytoplasm or approximately 700 molecules per cell on the surface. This threshold for detection is lower with the red-shifted variants.

GFP engineering is rapidly creating different colour mutants, from wild-type green to blue emitters and red-shifted excitation derivatives (*Table 4.2*). The blue-shifted GFP mutant, GFPuv, is 18 times brighter than wt GFP when excited by standard UV light due to the presence of amino acid mutations which increase the translational efficiency of the protein. GFPuv contains three amino acid substitutions (Phe-99 to Ser, Met-153 to Thr and Val-163 to Ala) resulting in an 18-fold increase in its fluorescence relative to wt GFP. A number of red-shifted excitation mutants fluoresce 4–35-fold more brightly than wt GFP when excited with blue light. The red-shifted GFP variants shift the maximal excitation peak to approximately 490 nm. One of the red-shifted mutants, P64L/S65T (enhanced GFP or EGFP), fluoresces 35-fold more intensely than wt GFP when excited at 488 nm due to an increase in its extinction coefficient. The extinction coefficient for EGFP has been measured at $250\,000$ cm^{-1} M^{-1} (488 nm excitation), compared with 7000 cm^{-1} M^{-1} for wt GFP under similar conditions. The coding sequence of this mutant

Table 4.2. Spectral properties of wild-type and mutant GFPs

Clone	Mutation	Excitation maximum (nm)	Emission maximum (nm)	Extinction coefficient ($M^{-1}cm^{-1}$)	Quantum yield	Comments
WT		395 (475)	508	21000 (7500)	0.77	
EGFP	P64L/S65T	488	507	250000		35× brighter than GFP
GFP405		405 (490)	525			Superbright
GFPA (Y66H)	Y66H/V163A/S175G	384	448			Thermotolerant
sg42	F64L/Y66H	384	450		0.24	
	F64L/Y66H/V163A	384	450		0.25	
GFPuv	F99S/M153T/V163A	395 (380)	509			
MmGFP		490	510			Thermotolerant, humanized, for transgenic animals
H9	S202F/T203I	398	511			
P4	Y66H	383	447	13500	0.21	
P4-1	S65T	489	511	39200	0.66	
P4-3	Y145F/Y66H	381	445	14500	0.38	
P9	I167V	471 (396)	502 (507)			
P11	I167T	471 (396)	502 (507)			
W2	Y66W/I123V/Y145H/H148R/H153T/V163A/N212K	432 (458)	480	10000 (9600)	0.72	
W7	Y66W/N146I M153T/V163A N212K	433 (453)	475 (501)	18000 (17100)	0.67	
	S65A	471	504			
	S65C	479	507			
	S65L	484	510			
	Y66F	360	442			
5B, 9B	T203H/S65T	512	524	19400		
6C	T203Y/S65T	513	525	14500		
10B	T203Y/F64L/S65G/S72A	513	525	30800		
10C	T203Y/S65G/V68L/S72A	513	527	36500	0.62	
11	T203W/S65G/S72A	502	512	33000		
Emerald	F64L/S65T/S72A/N149K/M153T/I167T/H231L	484	508	36000	0.7	Acceptor for EBFP
Topaz	S65G/S72A/K79R/T203Y/H231L	514	527	48000	0.63	Acceptor for CYP-W7
Sapphire	S72A/Y145F/T203I/H231L	395	511	33000	0.6	Donor for EGFP emerald

contains more than 190 silent base mutations that optimize the coding sequence for maximal translation efficiency in mammalian systems. Currently favoured filter sets for some of the GFPs are as follows: BFP – 390/22 ex; 420 dichroic; 460/50 em; EGP – 470/40 ex; 490 dichroic; 525/50 em; CFP(w7) – 440/20 ex; 455 dichroic; 480/35 em EYFP – 510/20

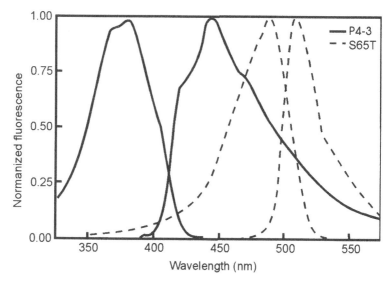

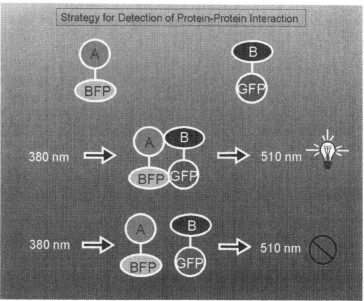

Figure 4.11. Excitation and emission spectrum (top) of P4-3 and S65T mutant green fluorescent proteins and strategy for detection of protein–protein interactions using FRET and mutant GFP proteins (bottom). If two proteins, one of which is labelled with P4-3 (BFP – donor) and the other with S65T (GFP – acceptor) physically interact, increased intensity at the acceptor emission maximum will be observed when the complex is excited at the wavelength of maximal absorbance of the donor.

ex; 520 dichroic; 530/25 em.

The availability of several different coloured mutants of GFP opens the attractive possibility of multiparameter and FRET imaging in living cells or tissues. Through the use of EBFP, CFP, EGFP and EYFP and judicious selection of appropriate excitation, dichroic and emission interference filters, it is possible to examine four distinct tagged cellular constituents in the same cell. Recently, preliminary experiments performed employing bright blue fluorescent protein (BFP, P4-3) and GFP (S65T) demonstrated the possibility of using these proteins for FRET experiments (*Figure 4.11*). BFP (P4-3) has an absorbance peak at 380 nm and an emission peak at 450 nm, while GFP (S65T) has an absorbance peak at 489 nm and emission peak at 511 nm. GFP (S65T) has an absorbance spectrum which overlaps with BFP's emission spectrum, providing a donor and acceptor pair for FRET experiments.

A recent example of the power of this approach is the construction of new fluorescent indicators for Ca^{2+}, termed 'chameleons'. Tsien and colleagues developed these indicators which consist of a fusion of mutant GFPs with calmodulin and the calmodulin-binding peptide M13. Two constructs were made. Blue or cyan mutant GFPs served as donors and green or yellow mutant GFPs served as acceptors. These constructs make measurements of Ca^{2+} in the range of 10^{-8} to 10^{-2} feasible.

However, the successful use of mutant GFPs for FRET microscopy requires that several points be considered. First, as stated previously, the donor and acceptor fluorophore concentrations need to be tightly controlled. This is currently a challenge when using GFP and BFP constructs, as each construct needs to be under the control of an inducible promoter, so that proper expression and appropriate donor/acceptor ratios can be obtained. In order to achieve high S/N, high fluorescence signals are preferred. This is currently a concern with BFP which has a low quantum yield. Conversely, too high a concentration of dye can cause self-quenching and disordered biological function. Second, photobleaching needs to be prevented. Almost all fluorescent molecules are sensitive to photobleaching, although the GFPs (but not the BFPs) are relatively resistant to it. Photobleaching can alter the donor/acceptor ratio and therefore the value of FRET. Lastly, GFP molecules tend to dimerize with high affinities and therefore it is necessary to determine whether any observed FRET is a consequence of true protein–protein interactions or an artifact of the propensity for GFP molecules to form dimers.

While GFP fluorescence is very stable and is more resistant to photobleaching than is fluorescein, there are a number of conditions that can enhance or destroy GFP fluorescence. GFP needs to be in an oxidized state to fluoresce, as chromophore formation is dependent upon an oxidation of Tyr-66. Strong reducing conditions (e.g. 5 mM Na_2SO_4 or 2 mM $FeSO_4$) can convert GFP into a non-fluorescent form. Weaker reducing agents, such as 2% β-mercaptoethanol, 10 mM dithiothreitol (DTT), 10 mM reduced glutathione, or 10 mM L-cysteine, and moderate oxidizing agents do not appear to affect the fluorescence of GFP. GFP fluorescence is irreversibly destroyed by 1% H_2O_2 and sulphydryl reagents such as 1 mM 5,5'-dithiobis(2-nitrobenzoic acid) (DTNB). GFP fluorescence is retained in mild denaturants, such as 1% sodium dodecyl sulphate (SDS) or 8 M urea, and after fixation with glutaraldehyde or formaldehyde, but is very sensitive to some nail polishes used to seal coverslips. GFP fluorescence is stable between pH 7 and 12, but is quenched at pH 5.5–7.0. GFP can dimerize via hydrophobic interactions at protein concentrations above 5–10 mg ml^{-1} and high salt concentrations, with a four-fold reduction in the absorption at 470 nm. GFP is exceptionally resistant to heat (melting temperature, $T_m = 70°C$), but the formation of the GFP chromophore is temperature sensitive. Stronger GFP fluorescence in bacteria has been found at 24 or 30°C compared to 37°C, and mammalian cells expressing GFP exhibit stronger fluorescence when grown at 30–33°C compared to 37°C. New mutant GFP molecules are being described at a rapid pace that alter the spectral and chemical properties of the GFP molecule.

4.2.1 Potential difficulties in using GFP fluorescence

A number of investigators have reported varying levels in the intensity of expressed GFP and GFP–protein constructs. This could be due to:

(i) The relatively slow formation of the GFP chromophore (this requires a period of hours).

(ii) The requirement for molecular oxygen (i.e. cells must be grown in an aerobic environment).

(iii) Expression of GFP below the limit of detection.

(iv) Failure of GFP to form the chromophore.

(v) Incorrect conditions for GFP expression (incorrect GFP plasmid construct; the vector being used is not compatible with your cell type; low transfection efficiency).

(vi) Use of an inappropriate filter set (filter sets commonly used for fluorescein isothiocyanate (FITC) fluorescence microscopy excite at 450–500 nm and can be used with EGFP; other narrower band-pass filter sets are commercially available which claim higher specificity but with lower sensitivity). Both Chroma and Omega manufacture numerous filter sets for use with GFP-expressing proteins.

(vii) Autofluorescence. Most autofluorescence in living mammalian cells is due to flavin coenzymes (FAD and FMN) which have absorption/emission = 450/515 nm. Since GFPuv uses excitation of 360–400 nm, autofluorescence of flavins should not be a problem. However, NADH, principally in mitochondria, does excite in this wavelength range and can contribute to autofluorescence.

In addition, certain types of growth media can cause autofluorescence. When possible, perform microscopy in a clear buffer such as PBS, or medium lacking Riboflavin and phenol red. Some cells produce a speckled autofluorescence pattern which is probably due to mitochondrially bound NADH. This problem is minimized with excitatory light around 488 nm, as opposed to UV excitation. Autofluorescence can also increase with time in culture. Always use a mock-transfected control or cells transfected with a promoterless vector such as pEGFP-1 or pGFP-1 to gauge the extent of autofluorescence. In fixed cells, autofluorescence can be reduced by washing with 0.1% sodium borohydride in PBS for 30 min after fixation.

5 Quantitative fluorescence microscopy

5.1 Absorbance versus fluorescence

When light of intensity I_0 enters a solution of fluorophore, a portion of the incident radiation is absorbed. The light which is not absorbed by the solution and which passes through the solution is denoted I. Thus, $(I_0 - I)$ represents the amount of light energy absorbed by the fluorophore solution, part of which will subsequently be released as fluorescence. As we learned in Chapter 1, the amount of fluorescence, F, is proportional to light absorbed ($= I_0 - I$) $\times$ Q (quantum yield of the fluorophore). This relationship can be written as:

$$F = Q(I_0 - I) \quad \text{[where } (I_0 - I) \approx I_0 \, \varepsilon[c]x].$$

The Lambert–Beer law states that absorption, A, is related to the incident intensity, I_0, and the emitted intensity, I, by:

$$A = -\log (I/I_0) \text{ or } I = I_0 e^{-A}.$$

The fluorescence intensity is then defined as:

$$F = Q(I_0 - I_0 e^{-A}) \text{ or } F = QI_0(1 - e^{-A}).$$

This simple relationship has fundamental importance in terms of quantitative fluorescence microscopy. From this relationship, it can be seen that when A approaches zero, e^{-A} approaches 1 and F approaches zero. Conversely, when A approaches infinity, e^{-A} approaches zero and $F = QI_0$. When A is very small, $1 - e^{-A}$ approaches A and $F = AQI_0$. The amount of light absorbed by the fluorophore solution will be related to the concentration of the fluorophore in the solution. Thus, $A \propto kc$, where c is the concentration and k is a constant. Substituting kc for A, at low fluorophore concentrations:

$$F = kcQI_0,$$

and thus the fluorescence intensity is directly proportional to the concentration of fluorophore. At high fluorophore concentration:

$$F = QI_0,$$

which is independent of fluorophore concentration. The practical implication of this is that the fluorescence intensity is directly related to the intensity of the excitation light, and for quantitative fluorescence measurements I_0 must be kept constant. In addition, the intensity of the fluorescence is directly proportional to the absorbance but only under conditions of low absorbance (i.e. F is proportional to the number of absorbing molecules).

It is instructive to examine what happens as the concentration of the absorbing species increases (*Figure 5.1*). As can be seen, when the sample absorbance is greater than 0.05 in a 1 cm pathlength, the linear relationship between absorbance and fluorescence emission found at low absorbance no longer holds, and at higher absorbance levels reduced fluorescence is observed.

Two phenomena are responsible for this situation. The first, known as the inner filter effect or excitation absorbance, is defined as a decrease in the amount of available excitation intensity in successive layers of the fluorescing object the further the layer is from the light source (*Figure 5.2*). This is a concentration-dependent phenomenon, happening with increased frequency as fluorophore concentration increases, and is due to high local absorption of the excitation light by the fluorophore. Another way to visualize this is that the layer nearest to the excitation light source receives more excitation light than the layers further away from the excitation light source.

There is a second reason for this deviation from linearity not related to Beer's law: emitted fluorescence light can be reabsorbed partially by surrounding fluorophores since most fluorophore absorption and emission spectra overlap to some extent. This causes a further reduction in fluorescence yield. This latter problem can be dealt with by using fluorophores whose excitation and absorption spectra do not overlap.

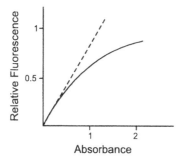

Figure 5.1. Relation between absorbance and fluorescence, calculated for a homogeneously stained object.

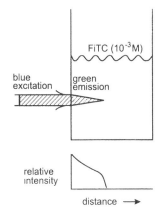

Figure 5.2. Inner filter effect, also known as excitation absorbance, is defined as a reduction of excitation intensity in a fluorescing object the further from the light source the fluorophores are. This is the result of high local absorbance. Not all fluorescent molecules absorb excitation light with the same efficiency. Those fluorophores nearest to the excitation light source receive more excitation light (I_0 is greater) than layers further from the excitation light source.

5.2   Calibration

Because fluorescence is linearly related to excitation intensity, fluorescence is a very sensitive and accurate way to measure concentration. However, fluorescence is a relative quantity, unlike absorbance (which is an absolute quantity), and a large number of factors can affect fluorescence emission, potentially affecting the quantitative accuracy of the fluorescent measurement. For example, it is not uncommon for the same specimen, when examined either using the same instrument on different days or with different instruments on the same day, to provide different levels of fluorescence. Therefore, it is important to be able to employ a standard preparation whose properties do not change and use this preparation to calibrate (normalize) the properties of the specimen emission.

There are two types of standard specimens used, instrumentation and preparation standards. Instrumentation standards are permanent specimens that fluoresce reproducibly under a defined set of circumstances. Historically, one such popular standard was a slide made of uranyl glass which gives off a strong yellow-green fluorescence. When the epi-objective is focused on the uranyl glass slide, the emission is essentially independent of the level of focus or position in the slide so long as the illuminated cone from the objective lies within the slide. However, uranyl glass is radioactive and thus fluorescent microspheres have replaced uranyl glass. Such standards are used to determine whether changes in the excitation intensity, sensitivity of the detector or electronic amplification occur during a given measurement (assuming the optical system, filters, diaphragms etc. remain unchanged).

Suspensions of microdroplets of known volume and fluorophore concentration in oil can also be employed as instrument standards. Spherical droplets of a fluorescent solution can be injected into immersion oil. After the droplets sink and come to rest (without flattening) on a siliconized coverslip, their diameter can be measured using an ocular micrometer, and their fluorescence measured with the same optics used to measure the real specimen. Droplet volume is then calculated from the droplet diameter.

Preparation standards are used to calibrate fluorescent staining reactions. These types of standards contain a known and/or constant amount of macromolecules that are fluorescently stained. Examples include diploid cells or chicken nucleated red blood cells as a reference for DNA measurements (both contain a 2C level of DNA), or sepharose beads with a known amount of fluorophore embedded in the bead matrix.

5.3   Instrumentation

Quantitative fluorescence microscopy can be performed using either transmitted or epi-illumination optical arrangements. In transmitted illumination, the excitation light is delivered from the bottom of the specimen, passing through the specimen and reaching the detector. This optical arrangement has been found to minimize reabsorption and provide better correlation between fluorescence intensity and fluorophore concentration. However, since the emitted fluorescence is much weaker than the intensity of excitation, it is critical to use filters capable of separating excitation light from the fluorescence emission. In epi-illumination, unwanted excitation light is removed by a dichromatic beam splitter in combination with a barrier filter. In addition, a second barrier filter of longer wavelength is often used to avoid the effect of reabsorption of the part of the fluorescence light that overlaps with the absorption spectrum on measured intensity.

Quantitative fluorescence microscopy can also be performed using either photomultiplier tubes (PMTs) or cameras as detectors. The benefits of these two types of devices are discussed in Chapters 2 and 6. PMTs are extremely sensitive, but lack the ability to provide spatial information at high temporal rates, while cameras can provide spatial information at high temporal rates, but are not as sensitive as PMTs. If one wishes to employ the sensitivity of the PMT and obtain spatial information, then scanning microscopes need to be employed.

5.4   Practical considerations

In Chapter 2, we discussed the components of a fluorescence microscope and a number of the particulars regarding optimization of fluorescence microscope performance. In quantitative fluorescence microscopy, all of these previously discussed optimizations apply, with the addition of the following important technical aspects. For optimal quantitative fluorescence microscopy, one requires high signal-to-noise. This is best accomplished by using high quality filters, objectives of moderate magnification and high NA, efficient light collection and sensitive photodetectors (which is preferable to having to employ large amounts of exciting light), and computer-controlled shutters to limit exposure of the specimen to damaging excitation energy. The number, wavelength-dependent transmission and quality of the optical elements of the microscope also dictate sensitivity. In addition, a standard fluorescing object which can be used for calibration and accurate measurement of the background light outside the fluorescing area is required, and the amount of photobleaching that occurs under the conditions of the experiment needs to be quantified.

It is absolutely essential that the light source energy output remains constant and stable during the measurement, as the fluorescence intensity is directly proportional to the exciting intensity. High pressure Hg or Xe lamps regulated by stable power supplies are normally employed. Insertion of a small photodiode into the excitation light path can be used to monitor lamp performance and can be integrated with a computer to provide continuous feedback to the lamp power supply to maintain lamp stability. In addition to lamp stability, the spectral and spatial distribution of illumination must be considered. Lamp spectral distribution will be dictated by the fluorophores to be employed, and placing a fibre-optic cable between the lamp and microscope will fully scramble any spatial variations of the light source. Non-uniformity of arc lamp epi-illumination over the field of view can also be corrected by dividing the specimen image on a pixel by pixel basis with a calibration image (which is an image of uniformly fluorescing material acquired under identical conditions as the specimen image).

The spectral and spatial sensitivity of the detector must also be characterized. PMTs, which measure photons falling on the photon-sensitive part of the PMT, have amplifiers to amplify the output signal and are sensitive to electronic noise as well as temperature. All PMTs display dark current (the signal obtained when no photons are hitting the PMT) and must be calibrated to zero before measurement. The PMT should demonstrate a linear response between output and exposure time assuming all other conditions remain constant. PMTs must be corrected for their different sensitivity at the different wavelengths of fluorescence emission.

The spectra of fluorophores can be influenced by the pH and ionic strength of the mounting medium. Therefore standardized preparation procedures must be employed for quantitative fluorescence microscopy. It also may be necessary to use an internal standard as a reference object. As we have previously discussed, the linear relationship between absorbance and fluorescence only holds at low absorbance (i.e. ≤ 0.2), so staining should be done in such a way as to keep the concentration of fluorophore low.

A common source of noise is non-specific fluorescence from slides, coverslips, embedding medium, immersion oil, objective lenses and auto-fluorescence. Since the contribution of these different components can vary with time, it is necessary to simultaneously determine both background and specimen-specific fluorescence intensities.

6 Digitized video microscopy

6.1 Digital imaging

Many of the specimens biologists examine contain low levels of fluorescence either due to low probe density in the sample or the need for minimization of excitation intensity to reduce fading or phototoxicity. Conventional wisdom states that the level of excitation intensity employed experimentally should be such as to minimize phototoxicity, while providing a detectable signal. A signal of 10^{-6} foot candles (fc) or 10 µlux (about 10% of what the dark-adapted eye can see) results in 4000 000 photons cm^{-2} sec^{-1} or less than 1 photon per pixel per video frame. Thus, use of the most sensitive fluorescence detection schemes is of the utmost importance.

Digital imaging is the process of converting optical information into numeric form, which allows the quantitative study of image data. Because this can be done for every point in the image at the same time, such a system gives the investigator both temporal and spatial photometric information at very high sampling rates. These rapid sampling rates allow extremely fast image processing operations to be performed. For example, images can be averaged as fast as they are acquired, producing an improvement in the signal-to-noise ratio of the final image which is proportional to the square root of the number of frames averaged. This is an extremely important tool to improve detection of weakly fluorescent specimens which are barely visible or invisible to the dark-adapted eye. Background sources of fluorescence from an unstained specimen can be stored and subtracted from the actual image to provide a more accurate representation of the distribution of the fluorescent label. Shading and geometric distortion can be removed and the resultant image 'grey stretched' to improve contrast. Finally, digital imaging allows for several standard ways of image display.

6.2 Digitized video microscopy (DVM)

DVM consists of capturing a fluorescent image produced by a standard fluorescence microscope using an ultra-sensitive camera, dividing the image up into discrete picture elements (pixels) and converting the intensity of fluorescence in each pixel to a number (digital value). The output from the camera is fed into an image processing computer which divides the image into an array of pixels (i.e. 512 × 512). The fluorescence intensity falling on each pixel is digitized to one of a number (i.e. 256) of grey levels within a frame acquisition time (30 msec). DVM techniques have found great utility in measuring the distribution and mobility of fluorescently labelled components on and within single living cells. To undertake such studies, one needs the ability to maximize the signal, keeping temporal resolution as high as possible, and yet not damage the cell. There are a number of ways one can maximize the signal and signal-to-noise in DVM. These include: (a) the use of probes with the highest quantum yields; (b) the use of detectors with the lowest dark counts and highest quantum efficiency; and (c) protection against photobleaching.

6.2.1 Imaging detectors

The imaging detector is the part of the digital imaging system where the majority of signal amplification occurs. Video is divided into two interlaced fields, displayed sequentially as odd and even numbered lines of the full frame. This ensures a field repetition rate of 50 or 60 Hz, slightly above the flicker fusion frequency of the human eye, thus giving a steady image despite the fact that the whole image is updated at only half that frequency. Standard video signals exist as complex waveforms that contain information about the brightness of each region of the display, along with horizontal and vertical sync timing pulses. A detailed discussion of video signals can be found in Inoué and Spring (1997).

The most commonly used broadcast-style video imaging devices are vidicon cameras. These cameras provide good image quality and moderately high horizontal resolution (limiting resolution of 600–1000 TV lines, depending on tube size and type). Some, such as Chalnicon or Newvicon tube cameras, are generally used for differential interference contrast (DIC) and bright-field microscopy. For microscopy involving very low light levels (10^{-4} to 10^{-8} fc), such as dark-field, fluorescence and high extinction polarized light microscopy of weakly retarding objects, vidicons may not be sensitive enough. Historically, the most widely used tube-based video cameras for fluorescence microscopy were the silicon intensified target (SIT) and intensified SIT (ISIT) because of their sensitivity (*Figure 6.1a*).

Ultra-sensitive, one-, two-, or three-stage multichannel plate (MCP) image intensifiers have further extended the ability to undertake low-light-level imaging (*Figure 6.1b*). An MCP is a secondary electron multiplier consisting of an array of millions of glass capillaries (channels). The internal diameter of each channel is 10 µm; channels are fused together to form a disc less than 1 mm thick. The inside wall of each channel is coated with a secondary electron emissive material and both ends are covered with a thin metal film (*Figure 6.1c*). Thus each channel becomes an independent secondary electron multiplier. When a voltage is applied between both sides of an MCP an electric field is generated in the direction of the channel axis. Secondary electrons, produced when an electron hits the entrance wall of the channel, are accelerated by the electric field and travel along parabolic trajectories determined by their initial velocity. They then strike the opposite wall and produce other secondary electrons. This process is repeated many times along the channel and, as a result, the electron current increases exponentially toward the output end of the channel (*Figure 6.1d*). MCP intensifiers are less noisy and less sensitive to blooming (oversaturation) than tube cameras. Proximity-focused intensifiers have virtually no shading or geometric distortion, and exhibit a lag less than 1 msec for a P-20 standard phosphor screen. An additional feature of proximity-focused tubes is their ability to be gated on a sub-nanosecond time-scale by changing or reversing the photocathode voltage. A single-stage MCP image intensifier intensifies light up to 4×10^4 times. For two- or three-stage MCP configurations, a gain of more than 10^5 can easily be attained. The intensified optical image formed on the phosphor screen is projected using a relay lens or fibre-optic coupling on to the target of a conventional high-resolution video camera or CCD, resulting in sensitivity two orders of magnitude greater than that of an SIT camera.

Unlike tube cameras, solid state CCD array detectors have the virtues of little geometric distortion, shading or lag (*Figure 6.2*). Three types of CCDs are used for quantitative electronic imaging. The classical full-frame CCD has a single parallel register used for photon collection, charge integration and charge transport. During the read-out process, the incident radiation must be blocked (by using an electronic shutter) in order to prevent distortion or smearing of the image. A frame-transfer CCD can operate continuously without a shutter at video frame rates. An opaque mask provides temporary storage for collected charge before read-out. The interline-transfer CCD has a parallel register that has been subdivided so that the opaque storage register fits between columns of pixels.

Frame-transfer devices operate by moving generated charge patterns very quickly from an image section to a storage section of equal size on the same chip. A line by line read-out from the storage area is accomplished while the next field of image charge is being collected (*Figure 6.2*). Both the interline- and frame-transfer array act in a parallel mode, compared to a full-frame-transfer array which acts in a serial mode. In

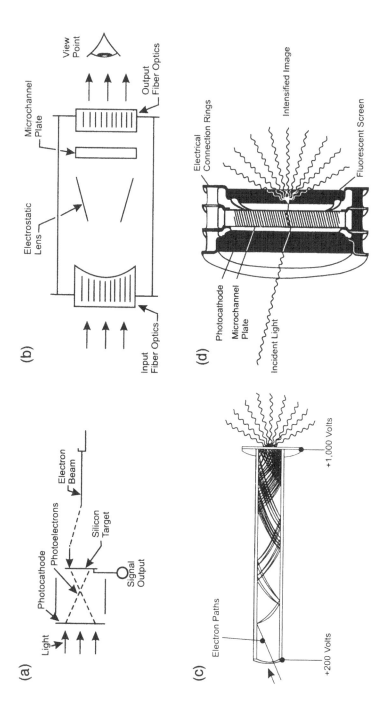

charge-injection device (CID) cameras, the X–Y addressed register systems usually use metal oxide semiconductor (MOS) capacitor pairs to store photo-generated charge. The signal is read sequentially by detecting voltage changes on the capacitance of the X and Y address registers. The CID has the ability to read data on the chip while the chip is acquiring information without destroying the information. When a CCD is read, the charge (i.e. the image data) accumulated on the chip is destroyed. CID cameras increase the dynamic range of the detector automatically, while minimizing analysis time. The dynamic range of these solid-state cameras covers the range from 5000:1 to 50 000:1, depending on the mode of operation.

SIT and CCD cameras are sensitive enough to detect low-light-level intensities (10^{-5} fc), but not sensitive enough to detect weak fluorescent signals, although the S/N of the CCD camera can be improved by gating the camera and allowing signal to accumulate (integrate) on the CCD chip before read-out. The ISIT is sensitive enough to detect low-light-level signals (as low as 10^{-8} fc) and the ISIT camera has been widely used as a low-light-level camera in fluorescence microscopy. The automatic gain mode in ISIT cameras provides an extraordinary operating range; however, gain reproducibility and stability may be poor under manual control.

Slow scan (0.1–4 Hz) cooled CCD array cameras enable prolonged integrating of exposures to detect low photon fluxes and reduce read-out noise. In these cameras the dark current noise is virtually eliminated and they have an extremely high signal-to-noise ratio or low noise (equivalent of 10 photons or less). The disadvantage of this camera is the slow read-out.

The use of colour solid state cameras in video microscopy, particularly for multiprobe labelling experiments, has increased due to improved sensitivity, resolution and colour reproduction. A cooled, colour, light-integrating CCD is a useful detector for fluorescence imaging. The

---

Figure 6.1. Schematic of Gen I (SIT) and Gen II (MCP) image intensifiers. (a) Gen I SIT intensifier. A photocathode converts incoming photons into photoelectrons, which are accelerated by a high voltage and impact upon a silicon target. The silicon target then generates a large number of electrons for each impinging electron, which are then detected by an electron beam sweeping the target. (b) Gen II MCP electrostatically focused intensifier. Photons enter the intensifier through a thick fibre-optic window that seals the evacuated interior of the image tube. On the inside of the window is the photocathode; the electrons it emits are focused by an electric field on to the microchannel plate. In this way, the brightened image is inverted, so that the image tube can be conveniently mated to additional optical systems that also invert the image. Hence the output image will not be inverted. By changing the focus of the electron beam the image can also be magnified or reduced in size. (c) Illustration of electron amplification in a single channel of a microchannel plate. (d) Proximity-focused intensifier. In a proximity-focused intensifier, the electrons emitted by the photocathode cross a short gap to the microchannel plate, where they are accelerated and multiplied. The electrons emerge from the plate and strike the anode, again across a short gap. The spatial organization of the input image is preserved because the two gaps are small enough to minimize electron dispersion. Adapted with permission of Plenum Press, New York, from Inoué, S. (1986) *Video Microscopy.*

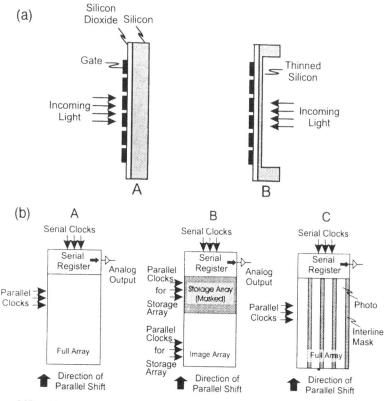

CCD architectures. (A) Full Frame CCD, (B) frame transfer CCD, (C) interline transfer CCD.

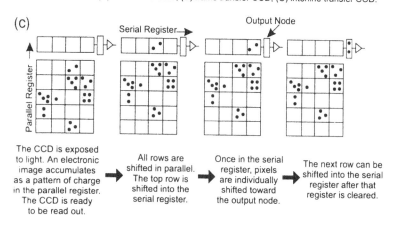

Figure 6.2. Charge coupled device (CCD) cameras: (a) thick (A) and thin (B) CCDs; (b) CCD architectures: (A) full frame CCD, (B) frame-transfer CCD, (C) interline-transfer CCD; (c) CCD read-out mechanism. Adapted with permission from K. Spring, Marine Biological Laboratories. Optical Microscopy and Imaging in the Biomedical Sciences, course notes.

colour camera generates images that are visually similar to what is seen directly through the eyepieces. Although use of a colour camera reduces the detection efficiency compared to a monochrome video camera, this can be compensated for by accumulating (integrating) the signal for longer times on the faceplate of the camera. The analogue outputs from the camera red, green and blue (RGB), or cyan, magenta and yellow (complementary) are first converted to hue, saturation and intensity (H, S and I). The colour analogue signals are then converted to digital images (using a colour digitizer) through an analogue-to-digital (A/D) converter. The H, S and I are represented in each pixel by three, 8-bit numbers. Generally, the arrangement of the colour filters on the surface of the imager is either in a stripe or mosaic pattern. In general, mosaic filters provide better horizontal resolution.

In many applications involving examination of fluorescence in living cells, the number of photons emitted by the fluorophores is limited, and intensified video cameras have enabled the detection, visualization and measurement of these low-light-level signals. In the ideal, the response of these low-light-level detectors should be a linear function of the flux of light reaching the camera's faceplate, providing a direct quantitative relationship between intensity recorded by the detector and fluorophore concentration. Unfortunately, this is not the case, and often the final image is corrupted to some extent by the detectors. Therefore, it is important to determine the performance of the detectors used in various biological digitized video microscopy applications. Cameras should be characterized in terms of temporal stability, shading, linearity, sensitivity and resolution.

6.2.2 Image acquisition and digital image processing

Acquisition of microscopic images and long-term storage are performed with different hardware depending on the application requirements. In studies of slow dynamic processes the output signal from the camera can be permanently archived on a time-lapse video cassette recorder (VCR) or an optical memory disk recorder (OMDR); the image can then be replayed in real time (which acts to 'electronically' speed up the process), photographed or printed as a hard copy.

The use of a VCR or OMDR provides a convenient, semi-quantitative mass storage medium. Quantitative image acquisition can be accomplished by digitizing the image using an A/D converter, a frame buffer and computer software and hardware. Computer processing can then be used to enhance the appearance of the microscope image, for example to increase contrast or to reduce noise, in ways that are difficult to do using photographic or video techniques alone.

Digitizers can either exist as stand-alone units or rely on the host computer central processing unit (CPU) to function. The host computer will influence the speed and versatility of the image processor as well as the type of operating environment the user will use. A stand-alone real-time digital image processor capable of operating on incoming video signals at 30 frames sec^{-1} should consist of the following: (i) A/D converter to transform incoming video to binary numbers; (ii) high-speed random access image memory for storage of digital images; (iii) arithmetic logic unit (ALU) which can perform numerical manipulations of images at video rates (add, subtract, divide); (iv) input and output look-up tables which can be used to alter the intensity values in the image according to pre-assigned functions; and (v) an output digital-to-analogue (D/A) converter to convert the digital image back to a video image. Additional specific hardware devices for performing convolutions, Fourier transforms and calculating histograms at video rates also exist.

The spatial resolution of a digitized image is related to the number of pixels (columns) per line and the number of lines (rows) into which the image is divided. Typical spatial resolutions are 512 pixels × 480 lines or 640 pixels × 480 lines. Brightness in a monochrome and colour image is dictated by the number of shades of grey and colours that can be distinguished, respectively. This is dictated by the A/D converter. A typical 8-bit A/D converter for a monochrome image represents an intensity spectrum, from total black to total white, of 256 shades of grey. In colour image systems, three A/D converters (8 bits each) simultaneously sample red, green and blue components of the colour signal (total of 24 bits, corresponding to 16 777 216 different colours). An average observer can discern thousands of shades and intensities in colour images but only one to two dozen shades of grey levels in monochrome images. The potential use of pseudocolour for monochrome image display and enhancement by assigning a colour to each pixel intensity helps detection and interpretation of information in the image.

Image processing software routines that perform image analysis of digitized microscopic images can vary greatly in their complexity and thus their requirements for a given type of hardware. Typical applications might be the determination of the number of cells in a microscope field of view, the measurement of nuclear diameter within a particular cell, or the measurement of fluorescent intensity in a particular area of the cell (e.g. nucleus or cytoplasm). Another example is automatic recognition and classification of normal and abnormal cells in a cytological smear. This might involve rapid measurement of many different image features, such as size and shape of individual cells, staining densities, and separation of overlapping cells using segmentation algorithms. Image enhancement routines can also be used to compensate for degradation of the original image caused by defects or limitations in the imaging system, such as aberrations in the optical system or non-uniform response to light of the video system.

6.3   Ion (ratio) imaging

One of the most popular applications of DVM is in the measurement of ion concentrations in the cytoplasm and organelles of single intact living cells. The use of fluorescent molecules to monitor ion concentration requires that a number of conditions be met.

(i) The indicator should have, if possible, a 1:1 stoichiometry with the ion being measured.

(ii) The measured signal is related to the ion being monitored. Often multiple ions can interact (albeit at different K_M values) with an individual ion indicator.

(iii) The optical response of the ion indicator must occur at the physiological concentration of the ion. This occurs best near the dissociation constant of the indicator. Thus, the dissociation constant needs to be matched to the intracellular concentration of the ion. Most indicators can report on ion concentrations over a 100-fold range surrounding the dissociation constant.

(iv) Changes which take place upon indicator–ion interaction can be multiple, and some are easier to measure than others. These changes include alterations in the quantum yield, a blue shift in the absorbance spectrum, or shifts in both the absorbance and emission spectra.

(v) As with other types of fluorescence microscopy, it is important to maximize signal-to-noise of the image.

(vi) Since these types of ion indicators work by binding the ion of interest, the concentration of the indicator must be such that it faithfully reports the concentration of the target ion, but does not buffer the target ion.

(vii) Intracellular localization of the dye is a prerequisite for proper interpretation of ion concentrations. There are a number of approaches which can be employed to load ion indicators into cells, including micro-injection, scrape-loading, fusion, electro- or chemical permeabilization, or by modification of the indicator into a form that will passively enter cells (e.g. acetoxymethyl (AM) esters). In the case of AM esters, intracellular esterases cleave off the AM group leaving a highly charged, membrane-impermeant form of the indicator in the cell. While this approach is convenient, it suffers from not being able to control the intracellular distribution of the indicator. In addition, not all cells may cleave off all of the AM esters on the indicator, leaving an indicator with altered sensitivity to its target ion. Lastly, cells can also pump out the ion indicator.

Historically, some of the earliest and most important ion indicators were those for measurement of intracellular Ca^{2+}. These compounds are fluorescent derivatives of the Ca^{2+} chelator *O,O'*-bis(2-aminophenyl)ethyleneglycol-*N,N,N',N'*-tetraacetic acid (BAPTA), an aromatic analogue of *O,O'*-bis(2-aminoethyl)ethyleneglycol-*N,N,N',N'*-tetraacetic acid (EGTA). Roger Tsien can be considered the father of the field of ion indicators, as he and his colleagues were the first to synthesize fluorescent Ca^{2+}-sensitive indicators. One of the first indicators they developed is known as fura-2 (*Figure 6.3*).

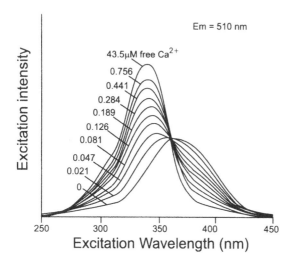

Em = 510 nm

43.5µM free Ca²⁺
0.756
0.441
0.284
0.189
0.126
0.081
0.047
0.021
0

Excitation intensity

250   300   350   400   450

Excitation Wavelength (nm)

Figure 6.3. Excitation spectrum of fura-2 as a function of Ca²⁺. Adapted with permission of Academic Press Ltd., London, from Mason, W.T. (1995) *Fluorescent and Luminescent Probes for Biological Activity.*

As can be seen, the excitation spectrum of fura-2 has two major peaks of absorbance (340 nm and 380 nm). As the level of Ca^{2+} changes, the absorption of light at these two wavelengths changes in opposite directions when the emission is recorded at 510 nm. As Ca^{2+} increases, the absorbance at 340 nm (and hence the emission at 510 nm) increases, while the absorbance at 380 nm (and hence the emission at 510 nm) decreases. The ratio of the emission intensity observed at the two different excitation wavelengths (340 nm and 380 nm) depends on the ion concentration but is independent of the concentration of the indicator. Fura-2 also has a wavelength at which no change in absorbance occurs as Ca^{2+} concentrations change. This is 362 nm and is known as the isosbestic point. For indicators undergoing a spectral shift upon ion binding, when the fluorescence is excited at the isosbestic wavelength the emission intensity is directly proportional to the dye concentration in the sample and independent of the concentrations of the ion.

The concept of ratioing indicator emission at two different wavelengths has been extended to DVM. Individual cells are not of uniform size; the perinuclear area is usually thicker than the very thin edge of the cell (i.e. the cell has a thicker/larger pathlength in regions near the nucleus than the edge of the cell). In addition, cells contain many organelles which also occupy volume in the cytoplasm and which can also change the effective pathlength of the cell. Due to the geometry of the light path in an epi-illumination system with respect to the sample, changes in intensity of an environmentally sensitive fluorophore could not only be due to a change in the concentration of the environmental factor that the fluorophore interacts with, but also to changes in pathlength or accessible volume. In a cell, because the excitation and emission arise from the same volume, ratio imaging normalizes for differences in optical pathlength, local probe concentration, and susceptibility of the probe to leak out of the cell or to undergo photobleaching (*Figure 6.4*). In cases where the ion indicator does not undergo a spec-

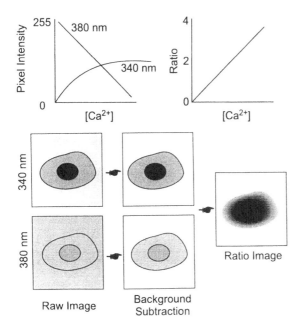

Figure 6.4. Principles of ratio imaging. In a cell, because the excitation and emission arise from the same volume, ratioing normalizes for differences in optical pathlength, local probe concentration, and susceptibility of the probe to leak out of the cell or to undergo photobleaching.

tral shift upon ion binding and/or lacks an isosbestic wavelength, it is possible to co-load cells with the ion indicator and another spectrally distinct fluorophore which is insensitive to ionic changes, and to ratio images obtained at the excitation and emission maxima of the two dyes.

6.4   Multiparameter digitized video microscopy (MDVM)

MDVM provides the ability to monitor a variety of cellular parameters in single living cells with quantitative temporal and spatial resolution. The very low light levels involved in this system minimize damage to the cells, and so permit the monitoring of cellular functions without disrupting normal physiology. Cultured cells or tissue are loaded with multiple probes whose fluorescence levels are specific for environmental parameters and whose excitation and/or emission are non-overlapping. By selecting excitation and emission wavelengths under computer control, MDVM will sample repeatedly over time the parameter-specific fluorescence of each of the probes. Thus, several different parameters can be followed simultaneously over time in single living cells. Since MDVM is an imaging technique, spatial patterns of these parameters within cells can also be determined.

6.5 Caged compound release/photoactivation

The recent availability of photolabile compounds, which upon photolysis with UV light release putative second messengers, provides the ability to study the role and regulation of agonist-induced alterations in cellu-

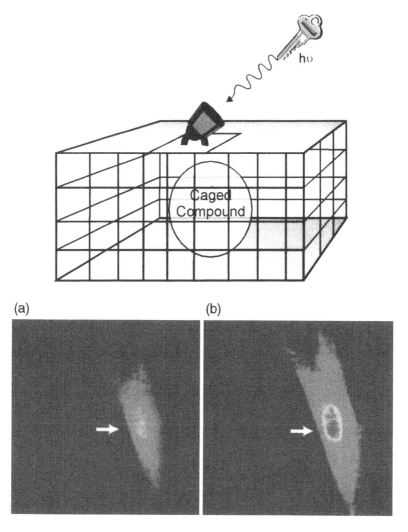

(a)   (b)

Figure 6.5. Caged compounds. Top: caged compounds are photosensitive precursors in which the presence of a photolabile masking group abolishes the bioactivity of an effector molecule until removal of the caging group by UV photolysis results in release of the bioactive molecule. Bottom: release of caged Ca^{2+} in nucleus of cell. A fibroblast was micro-injected with 20 mM Ca^{2+}-green dextran and 100 mM NITR-5 (caged Ca^{2+}). Thirty min after injection, resting Ca^{2+} levels were determined (a). The nucleus was then irradiated with an 800 msec pulse of 365 nm light and the Ca^{2+} levels determined by the change in Ca^{2+} green fluorescence (b).

lar second messengers involved in cellular physiology. The principle behind the use of caged compounds is based on the ability of UV irradiation to photolytically alter the chemical structure of the caged compound such that the binding and/or affinity of the overall structure for the 'caged' component of the complex is reduced, allowing its release from the caging structure (*Figure 6.5*). Photolytic conversion generally occurs with the redistribution of electronic charges in the compound, thereby reducing its affinity for the molecule it was harbouring and allowing its release from the caging structure. Once the caged material has been introduced into the cytoplasm of a cell (e.g. via micro-injection), it may be released from the cage with a pulse of UV light and subsequent changes in cellular activity (e.g. Ca^{2+}) can be monitored using MDVM in the same cell using a parameter-sensitive probe that does not absorb at the wavelength of light used for photolysis. This approach provides the investigator with the ability to modulate real-time cellular physiology in a spatial and temporal manner. A number of currently available caged compounds exist which can modulate intracellular Ca^{2+} levels, including IP_3, a non-hydrolysable analogue of IP_3 ($GPIP_2$), Ca^{2+} itself (NITR-7, NITR-5 or nitrophenyl-EGTA) and Ca^{2+} chelators (diazo-2 and caged BAPTA). Photoactivation is a specific type of caged compound release in which a fluorescent compound is rendered non-fluorescent due to chemical caging. When the caged fluorophore is irradiated with UV light, the endogenous fluorescence properties of the molecule can be observed.

7 Single and multiphoton microscopy

7.1 Principle of single versus multiphoton microscopy

Fluorescence is normally the result of absorption of an individual (single) photon of energy by a fluorescent molecule and the subsequent re-emission of some of that absorbed energy as light (*Figure 7.1*). For each fluorescent molecule, this process is repeated multiple times (e.g. for a single FITC molecule this is approx. 30 000 times), until excited state reactions completely bleach (destroy) the fluorophore. Multiphoton (two or three photon) absorption is the simultaneous (vs. sequential) absorption of multiple photons in a single quantized event that combines their excitation energies (in a virtual state) to produce an electronic excitation that is conventionally caused by a single photon of a shorter wavelength (e.g. two photons of red light can excite a UV-absorbing fluorophore).

Unlike single photon absorption, where the absorption efficiency is independent of the time of arrival of the exciting photon, multiphoton absorption probability is dictated by both spatial and temporal overlap of the incident photons. For two-photon excitation to occur, the two photons must arrive within 10^{-18} sec, resulting in an intermediate virtual state lifetime of 10^{-17} sec.

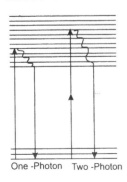

One -Photon   Two -Photon

Figure 7.1. Jablonski diagram showing excitation by one or two photons. For two-photon excitation, there is a short-lived virtual state that mediates the absorption (shown by the dashed line). Despite the fact that one- and two-photon absorption populate different energy levels in the excited state, internal conversion to the lowest energy level available causes the fluorescence emission characteristics to be independent of the excitation pathway. Adapted with permission of Wiley-Liss, New York, from Wang, X.F. and Herman, B. In: Winefordner, J.D., ed. (1996) *Fluorescence Imaging Spectroscopy and Microscopy Chemical Analaysis.*

7.2   Confocal microscopy

Conventional microscopes create images with a depth of field at high power of 2–3 μm but are theoretically capable of resolving structures as close as 0.2 μm (dependent on the wavelength of light). In a conventional epi-fluorescence microscope, the entire fluorescent specimen is exposed to excitation light and the emission collected by high NA objectives comes from throughout the specimen, including above and below the focal plane. Superimposition of detail within the plane of focus obscures structural detail that would otherwise be resolved. In addition, for specimens thicker than this depth of field, light from out-of-focus planes creates diffuse halos around objects of study. These halos are especially prominent in fluorescence microscopy. The result is that the image is seriously degraded due to reduced contrast and sharpness.

In contrast, confocal microscopes create optical sections which are approximately 0.5 μm thick. Confocal microscopy has gained acceptance as an important technology owing to its capability to produce images free of out-of-focus information. In confocal microscopy out-of-focus information (blur) is removed. Confocal microscopy rejects light from out-of-focus planes, and minimizes superimposition of detail, producing images which are remarkably detailed. Confocal microscopy can basically be thought of as a CAT scanner for cells. In addition, confocal microscopy provides a significant improvement in lateral resolution and the capacity for direct, non-invasive serial optical sectioning of intact, thick living specimens. All this improvement essentially comes from the rejection of out-of-focus interference. Increasingly, confocal microscopy has become an essential analytical tool to study the structure and physiology of living cells.

Confocal imaging systems remove out-of-focus information (intensity) by illuminating a single point of the specimen with a focused beam (so that illumination intensity drops off rapidly above and below the plane of focus), and by placing a pin-hole aperture in the emission light pathway at a focal plane conjugate to the specimen (*Figure 7.2*). A single diffraction-limited spot of light is projected on the specimen using a high NA objective lens and the light reflected or fluoresced by the specimen is collected by the objective and focused upon a pin-hole aperture. Light originating from above or below the image plane strikes the walls of the pin-hole and is not transmitted to the detector (PMT). If the pin-hole can be made small enough and the specimen is bright enough, confocal microscopy can increase the resolution that is obtained by a factor of up to 1.4 compared to conventional wide-field microscopy.

Confocal microscopy was introduced in 1957, and today most confocal microscopes are object plane scanners of two types: point scanning or slit scanning (*Figure 7.3*). This is in contrast to image plane scanners (i.e. cameras). Point-scanning confocal microscopes generally employ lasers

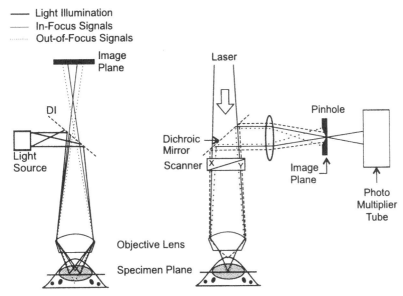

Figure 7.2. Comparison of the optical paths in wide-field and confocal epi-illumination fluorescence microscopes. In confocal microscopy, the excitation light is focused by the objective to a diffraction-limited spot at the focal plane of the specimen. Emitted fluorescence from both in- and out-of-focus planes is collected by the objective, but only the emitted fluorescence from the in-focus voxel is able to pass through the pin-hole aperture and can be detected by the PMT. Emitted fluorescence from planes above and below the in-focus plane are not focused at the aperture and therefore do not pass through the aperture.

for excitation and can detect (simultaneously) the emission of two or more fluorophores. For visible wavelength confocal microscopy, either an Ar ion laser that emits at 488 and 514 nm (which can excite fluorescein and rhodamine, respectively) or an Ar/Kr mixed gas laser (with emission at 488, 568 and 647 nm) is employed. Excitation of UV fluorophores has become possible recently with the advent of large frame, high power Ar ion lasers, whose output is from 351 to 360 nm.

Point-scanning confocal microscopes generally fall into two categories: stage scanning (SSCM) and laser scanning (LSCM). The SSCM employs a stationary laser as an excitation source, a photomultiplier as the detector and a specimen holder (stage) which moves and thus allows the specimen to be 'rapidly' scanned in the X–Y plane. A pin-hole in the emission path coupled with a high NA (1.4) objective lens removes out-of-focus information and sharply improves the contrast. This type of confocal microscope requires a relatively long period of time (approx. 10 sec) to acquire a single image, and is therefore limited to use with fixed specimens or microelectronic circuits, and is not usually used for live specimens where dynamic events are occurring.

In LSCM, generation of a two-dimensional image is accomplished by scanning the laser beam across the specimen (*Figure 7.3a*). The laser excitation is a diffraction-limited spot on the specimen resulting in rapid intensity fall-off above and below the plane of focus. The emitted

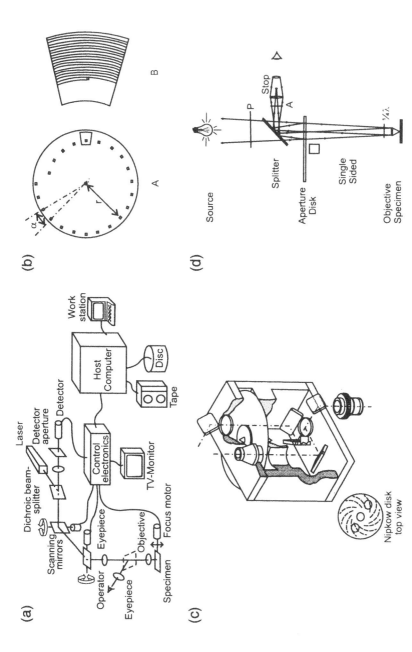

Figure 7.3. Schematics of various types of confocal microscopes. (a) Laser-scanning confocal microscope. (b) Nipkow disc. The perforations in the opaque disc (A) which is rotating at a constant velocity scan the image in a raster pattern as shown in part B. (c) Tandem scanning reflected light microscope. (d) Real-time scanning optical microscope. Adapted with permission of Plenum Press, New York, from Pauley, J.B. (1995) *Handbook of Confocal Microscopy*

fluorescence passes through a pin-hole aperture situated in a focal plane conjugate to the specimen. As previously described, this aperture will block any fluorescence originating from planes above or below this plane from reaching the detector. Most confocal imaging systems provide adjustable pin-hole blocking apertures. The optimum selection of pin-hole size is important, as the size of the pin-hole will dictate the intensity (brightness) and thickness of the slice observed. For instruments with variable pin-holes, an optimum pin-hole diameter should be determined empirically to provide the best combination of brightness and slice thickness. This enables a trade-off to be made in vertical resolution and sensitivity. A small pin-hole gives the highest resolution and lowest signal and vice versa. A computer is used to control the sequential scanning of the laser. To produce an image using LSCM, the laser beam must be moved in a regular two-dimensional raster scan across the specimen, and the instantaneous response of the photomultiplier must be displayed with equivalent spatial resolution and relative brightness at all points on the synchronously scanned phosphor screen of a CRT monitor.

The scanning is accomplished by galvanometer-driven mirrors, and the temporal resolution of a laser-scanning confocal microscope is dictated by the speed at which it can scan a focused spot of light across the specimen and, in doing so, generate an image with suitable signal-to-noise. For example, if we wish to collect an image consisting of 512×512 pixels in 1 sec, then the laser can only dwell at each pixel for 4 μsec. The limiting factor in temporal resolution is the fact that only a certain amount of fluorescence can be obtained from the small volume of fluorochrome contained within the excitation volume of the focused laser beam. Increasing excitation energy further will eventually result in all of the fluorophore molecules existing in the excited state (ground state depletion). The fluorophore is now 'saturated' as no more signal may be obtained from it by increasing the intensity of excitation source. However, if higher temporal resolution is required and two-dimensional spatial information can be sacrificed, a single line can be repetitively scanned across the field at intervals between 10 and 100 msec. Should higher temporal resolution be required, the line scanning component of the confocal could be deactivated and repetitive measurement of the same spot could be made repeatedly.

Point-scanning microscopes, when used with high NA lenses, have an inherent speed limitation in fluorescence. Most commercial scanning beam confocal microscopes have laser excitation sources that give around 10 mW in the major spectral lines. When the spectral line is near the excitation peak of the fluorochrome being used (e.g. the 488 nm Ar line and fluorescein) and a high NA lens is used (approx. 1.0 NA), this power level will cause saturation giving image degradation. Better images will be obtained by reducing the power by a factor of 10 or 100. This limits the speed which an image with a given signal-to-noise ratio can be acquired. Typically, about 5–10 sec of integration is required with an average immunofluorescence preparation.

Alternatively, improved temporal resolution can be achieved by exciting the specimen with an array of light beams in parallel, or a line of illumination which can be thought of as a one-dimensional array of points (*Figure 7.3b–d*). In scanning disc confocal microscopes, the pin-hole at the intermediate image plane is replaced by a perforated rotating disc containing pin-holes (Nipkow disc), so that the whole specimen is rapidly and continuously swept by an array of pin-holes. In a slit-scanning confocal microscope, the illuminating light source is scanned across the specimen through a narrow slit as opposed to a pin-hole aperture. The length of the slit is such that it covers the entire field of view of the objective lens, but the width of the slit is the diameter of the pin-hole that it replaced. The resultant emitted light is descanned and detected by a video camera after passage through a variable-aperture detection system. Slit-scanning confocal microscopes have a higher speed potential than point scanners because their use of an array of parallel beams avoids fluorophore saturation. Increases in beam scanning frequency have also been achieved using acousto-optical deflectors (AOD). An AOD is a non-mechanical device which increases beam scanning speed. In real-time AOD systems, the laser beam is rapidly scanned across the specimen by non-mechanical deflection through an arrangement of prisms and acousto-optical devices in two dimensions. The trade-off of increased time resolution, however, may be at the expense of a significant decrease in the signal-to-noise ratio of the output image. In addition, slit-scanning confocal systems do not **have** quite such good out-of-focus rejection as point-scanning systems. Some slit-scanning systems currently use a laser as a light source, while others use Hg or Xe lamps as light sources. The use of non-laser light sources allows for selection of any excitation wavelength and real-time imaging with direct viewing. An intensified video camera is usually used as the detector, providing a broader spectral response than is possible with PMTs.

7.3 Multiphoton microscopy

The goal of any type of fluorescence microscopy is to provide an image of suitable signal-to-noise without damage to the specimen. While confocal fluorescence microscopy substantially improves the spatial detail in images of 3D objects, it is still limited, especially when examining dynamic processes in living cells, due to phototoxicity and photobleaching of the fluorophore during observation. This effectively limits the excitation intensity that can be used and can result in low signal levels, precluding observation of dynamic events in cells and tissues. Remember, in confocal microscopy, the whole of the sample in the field of view is irradiated even though only one focal section (0.5 μm thickness) is being observed.

These problems can be minimized with two-photon excitation. Multiphoton microscopy is made possible by the very high local instantaneous intensity provided by a combination of diffraction-limited focusing of a single laser beam in the microscope and the temporal concentration of 100 fsec (femtosecond) pulses generated by a mode-locked laser. Using this approach results in peak excitation energies approximately 10^6 times greater than those used in confocal microscopy. However, the short duty cycle of the laser (approximately 10^{-5}) results in average power of approximately 1 mW, similar to that used in confocal microscopy. Since the excitation intensity varies as the square of the distance from the focal plane, two-photon absorption probability outside the focal region falls off with the fourth power of the distance along the optical axis (z direction). At 700 nm with an objective lens of 1.4 NA and a focal volume of 0.3 μm $\times$ 1.0 μm, 80% of the photon excitation will occur in the focal volume. Thus, 3D depth resolution is provided by the majority of the excitation being confined to a small part of the sample (the focal volume of excitation). In other words, fluorophore excitation will only occur at the point of focus (where it is needed), thus eliminating unnecessary photobleaching and phototoxicity.

As there is little excitation out of the plane of focus, image quality is excellent as there is practically no out-of-focus interference. The depth discrimination obtained is equal to that of a confocal microscope, but there is no out-of-focus photobleaching or photodamage, as is the case with confocal microscopy, and no attenuation of the excitation beam by out-of-focus absorption (*Figure 7.4*). Two-photon absorption also enhances the collection efficiency (due to the absence of a confocal pinhole aperture in the light path and the lack of a need to descan the emitted fluorescence). Two-photon excitation also allows excitation of UV

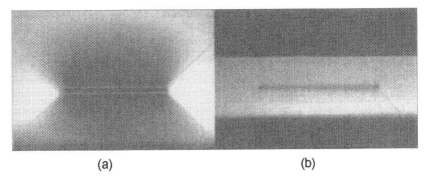

(a)   (b)

Figure 7.4. (a) *XZ* profile of the excitation pattern in a confocal microscope formed by repeatedly scanning a single *XY* optical section in a thick film of rhodamine formvar until the fluorescence was completely bleached. The box represents the region from which data were collected by the confocal microscope for this optical section. Nearly uniform bleaching occurred above and below the focal plane. (b) Same excitation pattern, but for two-photon excitation. The line denotes the path taken by the excitation light to reach the focal plane, but no excitation occurs outside the focal plane. Adapted with permission of Wiley-Liss, New York, from Wang, X.F. and Herman, B. In: Winefordner J.D. ed. (1996) *Fluorescence Imaging Spectroscopy and Microscopy Chemical Analysis.*

fluorophores with conventional visible light optics because both excitation and emission are in the visible parts of the spectrum.

The wavelength required for two-photon excitation is usually twice the wavelength required for one-photon excitation. (Likewise, the wavelength required for three-photon excitation is approximately three times the wavelength required for one photon excitation.) For example, in the case of fluorescein which has a one-photon absorption maximum at 493 nm, 986 nm excitation could be used for two-photon excitation. Three-photon excitation results in slightly improved z-axis resolution relative to two-photon excitation. Using three-photon excitation, it is possible to select fluorophores such that multiple labelled samples can be imaged by combination of two- and three-photon excitation, using a single IR excitation source. For example, three-photon absorption is advantageous because single photon UV excited fluorophores can be excited with IR excitation (e.g. indo-1 single-photon excitation is 338 nm; 1.014 µm excitation could be used for three-photon excitation).

In addition to these advantages, multiphoton excitation allows deeper penetration into tissue than confocal or wide-field microscopy. This is due to the fact that the excitation source is not attenuated by absorption by fluorescent molecules above the plane of focus, the longer excitation wavelengths used suffer less Raleigh scattering, and that the fluorescence signal is not degraded as much by scattering from within the sample. In addition, all the emitted photons from multiphoton excitation can be used for imaging because no confocal apertures have to be used. It is also possible to excite UV fluorophores using objective lenses that are not corrected for UV as UV wavelengths never have to pass through the lens.

7.4 Three-dimensional wide-field microscopy

We have just seen how it is possible to remove out-of-focus information from images using either one-photon confocal (by placing different types of mechanical apertures at proper positions in the microscope light path) or two-photon excitation microscopy. However, there is another approach that can be used to remove out-of-focus information from wide-field microscopic images. Three-dimensional data can be collected as a series of images (optical sectioning) obtained by sequentially incrementing the position of the focal plane (*Figure 7.5*).

Each of these images within the data set will contain not only in-focus information, but out-of-focus information as well. To remove the out-of-focus information from the image of interest, computer intensive image processing software is first used to provide an accurate description of the spatial degradation of light as it passes through the optical system of the microscope (point-spread function or optical-transfer function),

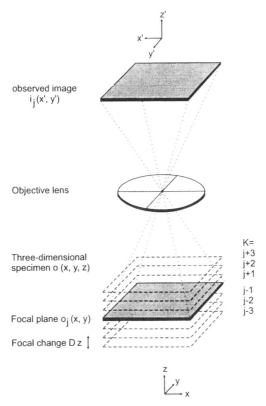

Figure 7.5. A schematic representation of the geometric relationship between a fluorescently labelled three-dimensional specimen and its observed blurred two-dimensional image. Each observed image, $i_j(x',y')$, is the sum of the in-focus information from the selected focal plane, $o_j(x,y)$, and the out-of-focus contributions from the remainder of the specimen, $o(x,y,z)$. Adapted with permission of Wiley-Liss, New York, from Shotton, D. (1993). *Electron Light Microscopy Techniques in Modern Biomedical Microscopy.*

and then the knowledge of the point-spread function or optical-transfer function is used to remove out-of-focus intensity from the image of interest. The point-spread function or optical-transfer function must be determined for each objective used in the microscope, as objective working distance and changes in the point-spread function as a function of depth can affect accuracy of the reconstruction. There exist a number of different algorithmic approaches for removal of out-of-focus fluorescence, including nearest neighbour deblurring, three-dimensional inverse filtering, constrained iterative deconvolution and EPR. Deconvolution techniques can also be used to further improve confocal images.

8 Fluorescence photomicroscopy

8.1 General principles

The goal of fluorescence photomicroscopy is to reproduce on to a permanent medium the image seen by the observer through the microscope as accurately as possible. Historically, film was the first medium employed to capture images through the microscope, and today still remains the preferred means of preserving specimen information. However, with the advent of electronic detectors and computer image processing hardware and software, other approaches can now be employed to preserve microscopic images with high fidelity. Fluorescence photomicroscopy is usually a race between integrating enough intensity on the storage medium (i.e. film) before the specimen is damaged (i.e. photobleaching). Thus, all of the principles we have talked about up to this point with respect to optimal operation of the microscope are even more critical for high quality photomicrographs.

8.2 Specimen preparation

Prior to mounting the specimen on a slide, it is necessary to wash the fluorescent specimens extensively to get rid of dried fluorochrome residues. The specimen slide should be clean (without dust or air bubbles), and one should use non-fluorescent microscope slides and cover-glasses, and non-fluorescent mounting media and immersion oil. The chemical environment of the fluorophore has a major impact on its emission properties (e.g. Q, photobleaching), and thus the mounting medium should be optimized for the fluorophores being examined. For fixed specimens, the use of chemical additives to minimize photobleaching and photodamage (e.g. addition of singlet oxygen scavengers) is strongly encouraged. Mounting medium of pH 8.5–9.0 has been found to max-imize the fluorescence of most common fluorophores.

8.3 Optimization of the microscope image

It is important to maximize the throughput of light from the excited specimen to the detector. Therefore, use of an objective lens of appropri-ate magnification (remember the intensity of the specimen decreases as the magnification increases; intensity is proportional to $1/mag^2$) which has the highest NA possible for the magnification and high transmission at the wavelengths to be employed, is essential. Epi-illumination should be used as this will allow better segregation of the exciting light from the emitted fluorescence. In addition, if possible 100% of emitted light should reach the detector. When employing a film camera, it is prefer-able to use the spot-metering mode of the camera. Full-field metering only provides correct values if the image is well mixed with grey values (i.e. phase contrast). When using an SLR-type camera, movement of the shutter and viewfinder mirror during exposure can cause vibrations in the camera/microscope that lead to blurring of the image. These prob-lems can be circumvented by using a remote-control system for camera operation and a vibration-free shutter. Selection of optimal exciter, dichromatic and emission filter wavelengths is also critical to the success of the instrument.

It is also possible to photograph still pictures off a video monitor; this is the most common means of documenting findings made via video (VCR) recordings. When making pictures directly from a TV-type moni-tor, the following guidelines should be employed:

- Use a clean monitor with the highest resolution possible. Clean the monitor with Kimwipes dampened with ethanol.
- Position the monitor flat on a table.
- Place the camera level and in front of the monitor, so the film plane is aligned parallel with the monitor face. Photograph the monitor in a darkened room or use a camera hood to prevent reflections and contrast loss of screen.
- Adjust the brightness and contrast slightly below your visual prefer-ence (final adjustments can be made when printing). Do not adjust the brightness too high as this will cause blooming and loss of detail.
- Adjust the vertical hold while looking through the camera viewfind-er. It should be adjusted to the point where the image becomes unstable and then reversed to just beyond where the image becomes stable again.
- Use a macro-lens (e.g. 55 mm focal length) and set the shutter speed to 1/8–1/2 sec with bracketing of the f-stop for proper exposure.

- Focus sharply on the scan lines of the raster. A Ronchi grating can be placed in front of the camera lens to de-emphasize the raster lines in printing.

8.4   Film

There exist two types of films for colour slides and for prints: daylight films and artificial light films. Daylight contains many more blue components than artificial light, which is mainly produced by glowing metal filaments and therefore tends towards yellow–red. When using colour film, it is important that the correct colour temperature be employed. Modern light microscopes feature the so-called 3200 K setting for the illuminator. This setting permits images on artificial light films to be produced without colour distortion. However, if you use a daylight film, the yellowish light from the microscope illuminator must be changed to bluish light. This is done by a colour conversion filter, which prevents a yellow tinge in the photo. Use daylight film unless blue light is being photographed. Use a tungsten balanced film for blue light.

Colour slide film is the most reliable medium for photomicroscopy. Kodak Ektachrome professional ASA 200, 400 or P800/1600 films are the best choices. Each of these films can be 'pushed'. Pushing a film means that it is exposed at a higher index (ASA or ISO setting) and is developed for a longer time. It is important to remember that any film when pushed produces higher contrast. Best image quality (resolving power and granularity) is always a trade-off with film speed (sensitivity). Using a film and processing that gives higher-speed film can result in grainy images and loss of specimen detail.

Film has a reduced efficiency when exposed to low light levels for long times (reciprocity failure). The reciprocity law of exposure states that if the total number of photons is the same, it matters not whether the exposure is of high intensity and short duration or of low intensity and long duration; the resultant images will have the same density. This holds for films between 1/1000 and 1 sec. Outside of this time-frame this law does not hold true. The practical implication of reciprocity failure is that if the specimen fluorescence is very weak, the film will be very inefficient. In a practical sense, this results in photobleaching of the specimen and incorrect colour reproduction. As previously discussed, this latter concern can be corrected for by employing colour compensating filters in the light path before the camera.

In selecting the appropriate film to use, the most important criterion is the ISO number which indicates the film speed. *Table 8.1* lists the differences in film sensitivity, grain resolution and exposure tolerance. Remember, other properties of the film material also change when the

Table 8.1. Film properties as a function of ISO number

ISO number	Sensitivity	Grain	Resolution	Exposure tolerance
25–50	Low	Fine	High	Wide
100–200	Medium	Medium	Medium	Medium
400+	High	Coarse	Low	Low

Adapted from Kapitza, H.G. (1996) *Microscopy From The Very Beginning.* Carl Zeiss, Jena.

speed changes. To find the optimum setting for the film it is necessary to perform a series of test images where the exposure time is prolonged or reduced.

8.5 The darkroom: film development and printing

Most black and white film is developed by the user in a darkroom. This allows substantially more flexibility to the user in terms of producing negatives and images of high resolution and contrast. The quality, contrast and grain of the film's emulsion is controlled by how the film is handled, the time of development, agitation and temperature. For black and white fluorescence pictures Kodak T-Max 400 with development extended by about 30% is a good choice.

It is very helpful to keep an ongoing accurate log of development conditions for both film and paper, as well as paper exposure times and enlarger lenses employed. In addition, distilled water should be used to make up all darkroom solutions, and the solutions should be stored in amber glass bottles and discarded when the chemicals have passed the manufacturer's rated usefulness. All developing trays and print tongs should be labelled and used for only one solution. After each darkroom use, they should be thoroughly rinsed. Filtering incoming air, if possible, will reduce dust, as will keeping the darkroom clean.

The general sequence for the development of film and paper is similar. Film is loaded in the dark into a film canister and sealed so as to prevent light from reaching the film, but allowing addition and removal of various solutions and washes. Once film is in the light-tight canister or the paper is exposed, it is placed in developer. Developers contain many ingredients, including agents that reduce light-exposed silver halides in the film emulsion to metallic silver, which is seen as the image. Sodium sulphite in the developer acts as a preservative by slowing oxidation of developing agents. Accelerators are alkalis that speed and strengthen the action of the developer. Restrainers (potassium bromide or benzotriazole) prevent unexposed silver halides from being developed. Following development, the film/paper is placed into a stop bath. The stop bath is a weak acetic acid solution which halts action of the developer. Following stop bath treatment, the film/paper is placed in fixer. Fixer (also known as hypo) is either sodium thiosulphate or ammonium thiosulphate with or without a hardener. The fix converts undeveloped silver halides into soluble salts that can be washed from the emulsion. The hardener (potassium alum) acts to strengthen the gelatin matrix in order to prevent excessive swelling or softening in the wash. Once fixed, the fixer is removed and the film/paper is rinsed in distilled water. Hypo clear, a salt solution which aids in the removal of the remaining fix and soluble silver complexes, is then added to the film/paper, and is followed by a final rinse in distilled water.

8.6   The digital darkroom: presentation of electronically (analogue or digital video) acquired data

Electronic photomicroscopy utilizes electronic devices (electronic detectors, computer hardware and software) to acquire, process, display and generate a hard copy of the original microscope image. In principle, this should be easy to accomplish, but distortion introduced by the electronic components (and, in cases when computers are used, hardware and software incompatibility) provides the greatest source of frustration in producing an accurate representation of the original microscope image.

The overall design of a typical digital darkroom is shown in *Figure 8.1*. A variety of different experimental technologies exist which generate data (e.g. microscope images, gels, slides, etc.). The devices used to generate either analogue or digital data sets of the original sample also vary substantially (e.g. cameras, VCRs, scanners, phosphoimagers, etc.). The use of a VCR allows easy copying, editing and narration of dynamic information captured through the microscope, which can also be distributed to others. However, it is important to use a VCR and tape that has sufficient lines of resolution, so as to retain important image detail, and which linearly (quantitatively) reproduces grey levels of the microscope image on to the tape. Often it is necessary to record a grey level wedge on to the video tape itself to allow accurate calibration.

A number of storage media exist which can capture microscopic images for storage and subsequent display in digital format. These include the optical memory disk recorder (OMDR), computer floppy, Zip, Jazz and CD ROM disks, as well as hard drives, laser disks and optical disks. It is very advantageous to be able to generate digital data sets for computer-based processing in a uniform format that is easily understood by the variety of image analysis/processing software currently available. The Tagged Image File Format (TIFF) is one such universal file format.

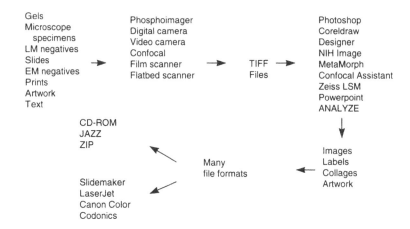

Figure 8.1. The digital darkroom. A variety of different types of data can be entered into the computer through various input devices. Subsequently, a number of software programs are used to prepare publication-quality images which are then used to make photographs or slides.

Once the images have been converted into a TIFF format, they can be processed to extract both qualitative and quantitative data, as well as prepare final images (i.e. collages) for publication. Commercially available image acquisition and display software often come equipped with a number of image processing options, including such manipulations as image averaging, background subtraction, ratio processing and a variety of graphical and pictorial representations of temporal and spatial image parameters. Unfortunately, different imaging systems and associated software do not necessarily employ a common or standard image format. For example, it is possible (actually probable) that the file format between the data acquisition and image presentation computers/software differs substantially. This often means that processing and file conversion software must be developed in-house if specific or specialized processing requirements cannot be met on the system on which the data was acquired. Image format capabilities also can become a concern in relation to the presentation of processed data in the form of pictures, slides or figures. For example, a number of graphic design programs exist which can be used for editing and presentation of microscope images. However, it may be the case that the input/output file formats used by these programs are not compatible either between programs or with specific hard copy (i.e. printer) output devices. As a general consideration, image processing software is most efficient and useful if it contains the flexibility of user-defined parameters and multiple data and file formats. Other important needs for software used for electronic photomicroscopy include bit map manipulation capabilities to allow adjustment of the printed colour image to the image seen on the computer monitor, the ability to prepare montages with labels, text, arrows etc. as necessary for publication or poster presentation, for printing colour images the ability to print in both CMYK and RGB formats, and that if the software uses postscript drivers for colour printing, ensuring that the postscript colour interpreter exploits the full ability of the printing objects.

A large number of choices exists for the production of a hard copy of microscope images. These choices range from photographic prints, colour copiers, a variety of types of printers, including ink jet which are becoming quite good and cost-effective, and dye sublimation which require special paper but have higher resolution and better colour rendition, and finally LCD projectors which allow display of computer acquired/processed images directly from the computer.

Appendix A

General optical system problems in fluorescence microscopy

TROUBLE	CAUSE	REMEDY
The lamp is operational but the image is very dark or unobservable	Shutter in illumination path closed; ND filter in use	Open shutter; remove ND filter
	The cube is not rotated in path correctly	Rotate cube into light path correctly
	Exciter and barrier filter combination incorrect	Use proper filter combinations for fluorochrome
	Aperture, field iris or objective iris diaphragm is not completely open	Open aperture, field iris or objective iris diaphragm completely
	Cube is unsuitable for use	Use suitable cube
	Lamp is not aligned properly	Align lamp
Unclear/blurred image, insufficient contrast	Dirty objectives or filters	Clean objectives/ filters
	Wrong combination of barrier and exciter filters	Use proper filter combinations for fluorochrome
	Aperture, field iris or objective iris diaphragm is not completely open	Open aperture, field iris or objective iris diaphragm completely

TROUBLE	CAUSE	REMEDY
	Cube is unsuitable for use	Use suitable cube
	No object in focus	Focus object using transmitted light
	Mirrors/reflectors out of position	Put mirrors/filters in proper position
	If microscope has attached camera, is light-directing prism set towards eye-pieces or camera?	Set light-directing prism correctly
	Lamp is not aligned properly	Align lamp
	Lamp output is weak	Check hours of lamp operation; change lamp if hours exceed average lifetime
Image partially obscured or unevenly illuminated	Extra (wrong) barrier filter objectives not inserted into light path correctly	Insert proper barrier filter Rotate nosepiece until it clicks
	Lamp is not aligned properly	Align lamp
	Filter cube is not rotated into light path correctly	Rotate filter cube into light path correctly
	Field iris diaphragm is closed	Open field iris diaphragm
	Shutter slider is not pushed in (open) enough	Push shutter slider all the way open
Excessive glare	Either exciter or barrier filter has not been inserted	Insert required filter

Adapted from Abramowitz, M. (1993) *Fluorescence Microscopy, The Essentials*, Vol. 4. Olympus-America, Inc. New York.

Appendix B

General electrical system problems in fluorescence microscopy

TROUBLE	CAUSE	REMEDY
Mercury or xenon lamp does not start		
Power switch indicator does not light up	Power cord not connected correctly	Connect correctly
	Bad fuses or faulty safety pin	Check fuses and safety pin
Power switch indicator light on but lamp does not start	Connectors are connected incorrectly; burner not installed	Connect correctly; install burner
	Lamp housing interlock is operating	Tighten bulb socket, locking screw securely
	Auto-ignition not operating	Turn off power supply. Try again
	Lamp used shortly before	Check to see if lamp housing is warm. If so, wait 10 min before restarting lamp
The bulb flickers or is dark	Insufficient time has elapsed since burner was turned on	Wait 10 min after turning on the burner

TROUBLE	**CAUSE**	**REMEDY**
	The bulb life has expired	Replace bulb if necessary. Remember the lamps are under pressure and present an explosion hazard. Wear safety glasses. Allow lamp to cool before replacing and never touch lamp with bare fingers.

Adapted from Abramowitz, M. (1993) *Fluorescence Microscopy, The Essentials*, Vol. 4. Olympus-America, Inc., New York.

Appendix C

Poor photographic results and their solutions

FAULT	SOURCE OF ERROR	REMEDY
Weak fluorescence, insufficient brightness	Specimens not stored properly, too old or bleached	Use fresh specimens or area of specimen that has not been bleached
	Rapid bleaching of specimens	Use transmitted light dark-ground or transmitted light phase contrast to select area of specimen; mount specimen in antifading medium
	NA of objective too small	Use objective with larger NA
	Excessive objective/ eyepiece magnification	Use low magnification eyepieces; low magnification/ high NA objectives
	Too small condenser aperture with transmitted light excitation	Use immersion dark-ground condenser. Apply non-fluorescent

FAULT	SOURCE OF ERROR	REMEDY
		immersion oil between front lens of condenser and underside of microscope slide
	Microscope room too bright	Darken room
	Not all light directed to camera	Set prism so all (100%) of emission light is directed to detector
	Film not sensitive enough	Use sensitive film (artificial light or tungsten-type colour films)
Picture is wrong colour	Improper colour filters	Use appropriate colour correction filters
	Film over-exposed	Do not over-expose film
	Blue background	From UV light; use correct barrier filter
Picture of poor contrast	Incorrect optics	Use optics with minimum magnification and maximum NA; close the field diaphragm slightly
	Excitation too broad-band	Use filters with narrower bandwidths
	Non-specific staining	Clean preparation; check purity of stain
	Fluorescent immersion oil	Use non-fluorescent oil
Autofluorescence or non-specific fluorescence	Contaminated preparation	Clean preparation
	Non-specific staining	Check specificity of stain through control; remove all unbound dye

FAULT	SOURCE OF ERROR	REMEDY
	Too broad-band excitation	Use filters with narrower bandwidths; use combination of filters; excite at lower wavelength than the optimum one using same barrier filter; use a barrier filter effective at higher wavelengths (sometimes a thicker filter is sufficient)
	Fluorescent immersion oil	Use non-fluorescent immersion oil
	Objectives unsuitable for fluorescence microscopy	Use only objectives recommended for fluorescence microscopy
With double fluorochroming, green and red picture details can be seen simultaneously	Use of filter block unsuitable for selective observation	Use narrower band filters that will allow selective observation of two fluorochromes individually
Sudden drop in fluorescence intensity	Specimen rapidly bleaches	Decrease excitation intensity; use neutral density filter; mount specimen in antifading medium
Light or red picture background	No red suppression filter in optical path	Put red suppression filter (SP610 or SP630) in optical path to block far-red emission light detected by the film; check that red-absorb-

FAULT	SOURCE OF ERROR	REMEDY
		ing BG38 filter is in illumination path
Uneven illumination	Poor lamp centration	Centre lamp
	Flickering lamp	Insert new lamp; check power supply
	With transmitted light excitation poor condenser centration	Centre dark-ground condenser; ensure oil contact with front lens of condenser and microscope slide
Impossible to focus microscopic image	Dirty front lens of the objective	Clean front lens of objective of immersion oil
Fine object contours appear unsharp	Flare	Use objectives with iris diaphragms

Adapted from Becker, E. (1989) *Fluorescence Microscopy, Principles, Instruments, Applications.* Wild Leitz GmbH, Wetzlar.

Appendix D

Difficulties in microfluorometry

TROUBLE	CAUSE	REMEDY
Strong background fluorescence	Detector dark current not set to zero	Set detector dark current to zero
	Excess room light	Darken the room
	Leakage of unwanted excitation light by filters	Use higher quality excitation filters
	Stray light	Set field diaphragm to minimize stray light
	Autofluorescence	Check immersion oil, filters and objectives
Variations in measured values	Photobleaching of sample	Mount specimen in antifading reagents; minimize intensity and duration of exposure during measurement
	Mismatch between field and measuring diaphragm	Centre field and measuring diaphragm and set to minimum size
	Lamp is unstable	Use stable lamp
	Non-specific dye	Make sure there is no free dye in sample

TROUBLE	CAUSE	REMEDY
	Improper lamp alignment	Align lamp in Köhler illumination
	Signal not in detector linear response range	Use neutral density filters; check linearity of detector and amplifier

Appendix E

Absorption and emission maxima for common fluorophores

FLUOROCHROME	ABSORPTION MAXIMUM (nm)	EMISSION MAXIMUM (nm)
Acid Fuchsin	540	630
Acridine Red	455–600	560–680
Acridine orange	500 (+DNA)	526 (+DNA)
	460 (+RNA)	650 (+RNA)
Acridine Yellow	470	550
Acriflavin	436	520
AFA (acriflavin Feulgen SITSA)	355–425	460
Alizarin Complexon	530–560	580
Alizarin red	330–560	580
Allophycocyanin	650	660
ACMA	430	474
AMCA-S, AMC	345	445
Aminoactinomycin D	555	655
7-Aminoactinomycin D-AAD	546	647
Aminocoumarin	350	445
Anthrovl stearate	361–381	446
Astrazon Brilliant Red 4C	500	585
Astrazon Orange R	470	540
Astrazon Red 6B	520	595
Astrazon Yellow 7 CLL	450	480
Atabrine	436	490
Auramine	460	550

FLUOROCHROME	ABSORPTION MAXIMUM (nm)	EMISSION MAXIMUM (nm)
Aurophosphine	450–490	515
Aurophosphine G	450	580
BAO 9 (bisaminophenyl-oxadiazole)	365	395
BCECF	482 (low pH)	520 (low pH)
	503 (high pH)	528 (high pH)
Berberine sulphate	430	550
Bisbenzamide	360	600–610
Blancophor FFG solution	390	470
Blancophor SV	370	435
BOBO-1, BO-PRO-1	462	481
BODIPY FL	505	513
BODIPY TMR	542	574
BODIPY TR	589	617
BOPRO 1	462	481
Brilliant sulphoflavin FF	430	520
Calcein	494	517
Calcein Blue	370	435
Calcium Crimson	590	615
Calcium Green	506	533
Calcium Orange	549	576
Calcofluor RW solution	370	440
Calcofluor White	440	500–520
Calcophor White ABT solution	380	475
Calcophor White standard solution	365	435
5- (and 6-)carboxy SNARF-1 indicator	548 (low pH)	587 (low pH)
	576 (high pH)	635 (high pH)
6-Carboxyrhodamine 6G	525	555
Cascade Blue	400	420
Catecholamine	410	470
CL-NERF	504 (low pH)	587 (low pH)
	514 (high pH)	540 (high pH)
Coriphosphine O	460	575
Coumarin-phalloidin	387	470
CY3.18	554	568
CY5.18	649	666
CY7	710	805
DANS (1-dimethyl amino naphaline-5-sulphonic acid)	340	525
Dansyl NH-CH$_3$ in water	340	578
DAPI	358	461

FLUOROCHROME	ABSORPTION MAXIMUM (NM)	EMISSION MAXIMUM (NM)
DiA	456	590
Diamino phenyl oxydiazole (DAO)	280	460
Di-8-ANEPPS	488	605
DiD [DiIC$_{18}$(5)]	644	665
Dimethylamino-5-sulphonic acid	310–370	520
DiI [DiIC$_{18}$(3)]	549	565
DiO [DiOC$_{18}$(3)]	484	501
Diphenyl brilliant flavine 7GFF	430	520
DM-NERF	497 (low pH) 510 (high pH)	527 (low pH) 536 (high pH)
Dopamine	340	490–520
ELF-97 alcohol	345	530
Eosin	524	544
Erythrosin ITC	530	558
Ethidium bromide	518	605
Euchrysin	430	540
FIF (formaldehyde-induced fluorescence)	405	433
Flazo Orange	375–530	612
Fluorescein	494	518
Fluorescein isothiocyanate (FITC)	490	523
Fluo-3	506	526
FM1-43	479	598
Fura-2	363 (low [Ca^{2+}]) 335 (high [Ca^{2+}])	512 (low [Ca^{2+}]) 505 (high [Ca^{2+}])
Fura Red	472 (low [Ca^{2+}]) 436 (high [Ca^{2+}])	657 (low [Ca^{2+}]) 637 (high [Ca^{2+}])
Genacryl Brilliant Red B	520	590
Genacryl Brilliant Yellow 1 OGF	430	485
Genacryl Pink 3G	470	583
Genacryl Yellow 5GF	430	475
Gloxalic Acid	405	460
Granular Blue	355	425
Haematoporphyrin	530–560	580
Hoechst 33258, Hoechst 33342	352	461
3-Hydroxypyrene-5,8,10-tri-sulphonic acid	403	513
7-Hydroxy-4-methylcoumarin	360	455
5-Hydroxytryptamine (5-HT)	400	530

FLUOROCHROME	ABSORPTION MAXIMUM (nm)	EMISSION MAXIMUM (nm)
Intrawhite cf liquid	360	430
Indo-1	346 (low [Ca^{2+}])	475 (low [Ca^{2+}])
	330 (high [Ca^{2+}])	401 (high [Ca^{2+}])
Leucophor PAF	370	430
Leucophor SF	380	465
Leucophor WS	395	465
Lissamine rhodamine B	570	590
Lucifer yellow CH	428	536
Lucifer Yellow VS	430	535
LysoSensor Blue DND-192, DND-167	374	425
LysoSensor Green DND-153, DND-189	442	505
LysoSensor Yellow/Blue	384 (low pH)	540 (low pH)
	329 (high pH)	440 (high pH)
LysoTracker Green	504	511
LysoTracker Yellow	534	551
LysoTracker Red	577	592
Magdala Red	524	600
Magnesium Green	506	531
Magnesium Orange	550	575
Maxilon Brilliant Flavin 10 GFF	450	495
Maxilon Brilliant Flavin 8 GFF	460	495
Mitotracker Green FM	490	516
Mitotracker Orange CMTMRos	551	576
MPS (methyl Green Pyronine stilbene)	364	395
Mithramycin	450	570
NBD	465	535
NBD amine	450	530
Nitrobenzoxadidole	460–470	510–650
Noradrenaline	340	490–520
Nuclear Fast Red	289–530	580
Nuclear Yellow	365	495
Nylosan Brilliant Flavin E8G	460	510
Oregon Green 488 fluorophore	496	524
Oregon Green 500 fluorophore	503	522
Oregon Green 514 fluorophore	511	530

FLUOROCHROME	ABSORPTION MAXIMUM (nm)	EMISSION MAXIMUM (nm)
Pararosaniline (Feulgen)	570	625
Phorwite AR solution	360	430
Phorwite BKL	370	430
Phorwite Rev	380	430
Phorwite RPA	375	430
Phosphine 3R	465	565
Phycoerythrin R	480–565	578
Pontochrome Blue Black	535–553	605
POPO-1, PO-PRO-1	434	456
Primuline	410	550
Procion Yellow	470	600
Propidium iodide	536	617
Pyronine	410	540
Pyronine B	540–590	560–650
Pyrozal Brilliant Flavin 7GF	365	495
Quinacrine Mustard	423	503
R-phycoerythrin	565	575
Resorufin	570	585
RH 414	500	635
Rhodamine 110	496	520
Rhodamine 123	507	529
Rhodamine 5 GLD	470	565
Rhodamine 6G	526	555
Rhodamine B	540	625
Rhodamine B 200	523–557	595
Rhodamine B extra	550	605
Rhodamine BB	540	580
Rhodamine BG	540	572
Rhodamine Green fluorophore	502	527
Rhodamine Red	570	590
Rhodamine WT	530	555
Rhodol Green fluorophore	499	525
Rose Bengal	540	550–600
Serotonin	365	520–540
Sevron Brilliant Red 2B	520	595
Sevron Brilliant Red 4G	500	583
Sevron Brilliant Red B	530	590
Sevron Orange	440	530
Sevron Yellow L	430	490
SITS (Primuline)	395–425	450
SITS (stilbene isothiosulphonic acid)	365	460
Sodium green	507	535

FLUOROCHROME	ABSORPTION MAXIMUM (nm)	EMISSION MAXIMUM (nm)
Stilbene	335	440
SNARF-1	563	639
Sulphorhodamine B can C	520	595
Sulphorhodamine G Extra	470	570
SYTOX Green nucleic acid stain	504	523
SYTO Green fluorescent nucleic acid stains	494 ± 6	515 ±7
SYTO Green fluorescent nucleic acid stains	515 ± 7	543 ± 13
SYTO 17 red fluorescent nucleic acid stain	621	634
Tetracycline	390	560
Tetramethylrhodamine, Rhodamine B	555	580
Texas Red	595	615
Thiazine Red R	510	580
Thioflavin S	430	550
Thioflavin TCN	350	460
Thioflavin 5	430	550
Thiolyte	370–385	477–484
Thiozol Orange	453	480
Tinopol CBS	390	430
TOTO-1, TO-PRO-1	514	533
TOTO-3, TO-PRO-3	642	660
True Blue	365	420–430
Ultralite	656	678
Uranine B	420	520
Uvitex SFC	365	435
X-Rhodamine	580	605
Xylene Orange	546	580
XRITC	582	601
YOYO-1,YOYO-PRO-1	491	509
YOYO-3,YOYO-PRO-3	612	613

Adapted with permission of Molecular Probes, Inc., Eugene, Oregon, from Haugland, R.P. (1996) *Handbook of Fluorescent Probes and Research Chemicals*, 6th Edn.

Appendix F

Further reading

Useful textbooks/articles on fluorescence and fluorescence microscopy

Abramowitz, M. (1993) *Fluorescence Microscopy – The Essentials*. Olympus-America, New York.

Allen, R.D., Allen, N.S. and Travis, J.L. (1981) Video-enhanced contrast, differential interference contrast (AVEC-DIC) microscopy: a new method capable of analyzing microtubule-related motility in the reticulopodial network of *Allogromia laticollaris*. *Cell Motil.* **1**: 291–302.

Andrews, C.L. (1990) *Optics of the Electromagnetic Spectrum*. Prentice-Hall, Englewood Cliffs, NJ.

Barer, R. and Coslett, V.E. (1996) *Advances in Optical and Electron Microscopy*, Vols I, II and III. Academic Press, San Diego, CA.

Becker, E. (1989) *Fluorescence Microscopy*. Wild Leitz, Wetzlar.

Bereiter-Hahn, J., Fox, C.H. and Thorel, B. (1979) Quantitative reflection contrast microscopy of living cells. *J. Cell Biol.* **82**: 767–779.

Block, S.M. (1990) Optical tweezers: a new tool for biophysics. In *Noninvasive Techniques in Cell Biology*, pp. 375–402. Wiley-Liss, New York.

Born, M. and Wolf, E. (1967) *Principles of Optics*. Pergamon Press, Oxford.

Bradbury, S. and Turner, G.L.E. (1967) *Historical Aspects of Microscopy*. W. Heffer and Sons, Cambridge.

Bradbury, S. (1967) *The Evolution of the Microscope*. Pergamon Press, Oxford.

Castleman, K.R. (1979) *Digital Image Processing*. Prentice Hall, Upper Saddle River, NJ.

Cheng, P.C., Wu, W.L., Lin, T.H. and Wu, J.L. (1994) *Multidimensional Microscopy*. Springer Verlag, Heidelberg.

123

Clark, G.L. (1961) *The Encyclopedia of Microscopy*. Reinhold, New York.

Clay, R.S. and Court, T.H. (1932) *The History of the Microscope*. Chas. Griffin and Co., London.

Delly, J.G. (1980) Photography through the microscope. *Kodak Publication P-2*.

DeWeer, P. and Salzberg, B.M. (eds) (1986) *Optical Methods in Cell Physiology*. Wiley-Liss, New York.

Dewey, T.G. (1991) *Biophysical and Biochemical Aspects of Fluorescence*. Plenum Press, New York.

Ditchburn, R.W. (1963) *Light*. Wiley Interscience, New York.

Francon, M. (1961) *Progress in Microscopy*. Pergamon Press, Oxford.

Francon, M. (1966) *Diffraction Coherence in Optics*. Pergamon Press, Oxford.

Francon, M. (1966) *Optical Interferometry*. Academic Press, New York.

Gonzales, R.C. and Wintz, P. (1987) *Digital Image Processing*, 2nd Edn. Addison-Wesley, Reading, MA.

Gregory, S.S. (1977) *Fluorescence and Phosphorescence Spectroscopy: Physiochemical Principles and Practice*. Pergamon Press, Oxford.

Gu, M. (1996) *Principles of Three-Dimensional Imaging in Confocal Microscopes*. World Scientific, Singapore.

Guilbault, G.G. (1973) *Practical Fluorescence – Theory, Methods and Techniques*. Marcel Dekker, New York.

Hale, A.J. (1958) *The Interference Microscope in Biological Research*. S. Livingston, Edinburgh.

Harris, D.A. and Bashford, C.L. (eds) (1987) *Spectrophotometry and Spectrofluorimetry: A Practical Approach*. IRL Press, Oxford.

Hartley, W.J. (1964) *How to use a Microscope*. Doubleday, New York.

Hecht, E. (1987) *Optics*, 2nd Edn. Addison-Wesley, Reading, MA.

Hemmilä, I.A. (1991) *Applications of Fluorescence in Immunoassays*. John Wiley and Sons, New York.

Herman, B. and Lemasters, J.J. (1993) *Optical Microscopy: Emerging Methods and Applications*. Academic Press, San Diego, CA.

Herman, B. (1998) *Fluorescence Microscopy*. Bios Scientific Publishers, Oxford.

Herman, B. and Jacobson, K. (1990) *Optical Microscopy for Biology*. Wiley-Liss, New York.

Hoffman, R. (1977) The modulation contrast microscope: principles and performance. *J. Micros.* **110**: 205–222.

Hoffman, R. and Gross, L. (1975) The modulation contrast microscope. *Nature* **254**: 586–588.

Hoffman, R. and Gross, L. (1975) Modulation contrast microscopy. *Appl. Optics* **14**: 1169–1176.

Inoué, S. (1987) Video microscopy of living cells and dynamic molecular assemblies. *Appl. Optics* **26**: 3219–3225.

Inoué, S. (1988) Progress in video microscopy. *Cell Motil. Cytoskel.* **10**: 13–17.

Inoué, S. and Spring, K. (1997) *Video Microscopy: The Fundamentals.* Plenum Press, New York.

Ito, H., Uede, T., Sugiyama, T., Iwasaki, R., Narimatsu, E., Yachi, A. and Kikuchi, K.J. (1993) The distinction of eosinophils from other hematopoietic cells by darkfield illumination. *Biol. Photog.* **61**: 141–143.

James, J. and Tanke, H.J. (1987) *Biomedical Light Microscopy.* Kluwer Academic, The Netherlands.

Jameson, D.M. and Reinhart, G.D. (eds) (1989) *Fluorescent Biomolecules.* Plenum Press, New York.

Jenkins, F.A. and White, H.E. (1976) *Fundamentals of Optics*, 4th Edn. McGraw-Hill, New York.

Kapitza, H.G. (1996) *Microscopy From the Very Beginning.* Carl Zeiss, Jena.

Kotyk, A. and Slavick, J. (1989) *Intracellular pH and its Measurement.* CRC Press, Boca Raton, FL.

Kriete, A. (1992) *Visualization in Biomedical Microscopies: 3-D Imaging and Computer Applications.* VCH Publishing, Weinheim.

Lacey, A.J. (1989) *Light Microscopy in Biology.* IRL Press, Oxford.

Lakowicz, J.R. (1983) *Principles of Fluorescence Spectroscopy.* Plenum Press, New York.

Lakowicz, J.R. (ed.) (1997) *Topics in Fluorescence Spectroscopy: Techniques (Volume 1); Principles (Volume 2); Biochemical Applications (Volume 3).* Plenum Press, New York.

Loveland, R.P. (1981) *Photomicrography: A Comprehensive Treatise*, 2nd Edn, Vols 1 and 2. Wiley, New York.

Marmasse, C. (1980) *Microscopes and Their Uses.* Gordon and Breach, Oxon.

Mason, W.T. (1993) *Fluorescent and Luminescent Probes for Biological Activity.* Academic Press, London.

Mollring, F.K. (1981) *Microscopy from the Very Beginning.* Carl Zeiss, New York.

Nuccitelli, R. and Deamer, D.W. (1982) *Intracellular pH – Its Measurement, Regulation, and Utilization in Cellular Functions*, Kroc Foundation Series, Vol. 15. Alan R. Liss, New York.

Nuccitelli, R. (ed.) (1994) *A Practical Guide to the Study of Calcium in Living Cells. Methods in Cell Biology*, Vol. 40. Academic Press, New York.

Parker, C.A. (1968) *Photoluminescence of Solutions.* Elsevier, Oxford.

Patzelt, W.J. (1985) *Polarized Light Microscopy: Principles, Instruments, Applications*, 3rd Edn. E. Leitz, Wetzlar.

Pawley, J. (1995) *Handbook of Biological Confocal Microscopy*, 2nd Edn. Plenum Press, New York.

Payne, B.O. (1957) *Microscope Design and Construction.* Cooke, Troughton and Sims, York.

Pesce, A.J., Rosen, C.-G. and Pasby, T.L. (1971) *Fluorescence Spectroscopy: An Introduction for Biology and Medicine.* Marcel Dekker, New York.

Piller, H. (1977) *Microscope Photometry*, Springer-Verlag, Berlin.

Ploem, J.S. and Tanke, H.J. (1987) *Introduction to Fluorescence Microscopy*. Oxford University Press, Oxford.

Pluta, M. (1989) *Advanced Light Microscopy*, Vols 1, 2 and 3. Elsevier, New York.

Reichman, J. (1994) *Glossary of Related Terminology with Page References to the Handbook of Optical Filters for Fluorescence Microscopy*. Chroma Technology Corp., Brattleboro, VT.

Reichman, J. (1994) *Handbook of Optical Filters for Fluorescence Microscopy*. Chroma Technology Corp., Brattleboro, VT.

Rendell, D. (1987) *Fluorescence and Phosphorescence*. John Wiley and Sons, New York.

Richardson, J.H. (1991) *Handbook for the Light Microscope: A User's Guide*. Noyes, New York.

Ross, K.F.A. (1967) *Phase Contrast and Interference Microscopy for Cell Biologists*. St Martin's Press, New York.

Rost, F.W.D. (1991) *Quantitative Fluorescence Microscopy*. Cambridge University Press, Cambridge.

Rost, F.W.D. (1992) *Fluorescence Microscopy*, Vols 1 and 2. Cambridge University Press, Cambridge.

Rubbi, C.P. (1994) *Light Microscopy: Essential Data*. John Wiley and Sons, New York.

Russ, J.C. (1994) *The Image Processing Handbook*, 2nd Edn. CRC Press, London.

Salmon, E.D., Inoue, T. and Murray, A.W. (1994) High resolution multimode digital imaging system for mitosis studies *in vivo* and *in vitro*. *Biol. Bull.* **187**: 231–232.

Salmon, E.D. (1995) Video-enhanced differential interference contrast (VE-DIC) light microscopy and the discovery of kinesin. *Trends Cell Biol.* **5**: 154–158.

Sharif, N.A. (1993) *Molecular Imaging in Neuroscience: A Practical Approach*. IRL Press, Oxford.

Shotten, D. (1993) *Electronic Light Microscopy*. Wiley-Liss, New York.

Shurcliffe, W.A. (1962) *Polarized Light*. Harvard Press, Cambridge, MA.

Shurcliffe, W.A. and Ballard, S.S. (1964) *Polarized Light*. Van Nostrand, Princeton, NJ.

Slavik, J. (1994) *Fluorescent Probes in Cellular and Molecular Biology*. CRC Press, Boca Raton, FL.

Slavik, J. (ed.) (1996) *Fluorescence Microscopy and Fluorescent Probes*. Plenum Press, New York.

Slayter, E.M. (1970) *Optical Methods in Biology*. Wiley Interscience, New York.

Slayter, E.M. and Slayter, H.S. (1992) *Light and Electron Microscopy*. Cambridge University Press, Cambridge.

Smith, R.F. (1994) *Microscopy and Photomicrography: A Working Manual*. CRC/Lewis Publishers, Boca Raton, FL.

Spencer, M. (1982) *Fundamentals of Light Microscopy.* Cambridge University Press, Cambridge.

Stevens, J.K., Mills, L.R. and Trogadis, J.E. (eds) (1994) *Three-Dimensional Confocal Microscopy: Volume Investigation of Biological Specimens (Cell Biology).* Academic Press, San Diego, CA.

Swindell, W. (1975) *Polarized Light,* Benchmark Papers in Optics. Halsted Press/Wiley, New York.

Tanke, H.J. (1995) *Fluorescence Microscopy* (Microscope Handbooks, No. 35). Coronet Books, Oxfordshire.

Taylor, D.L. and Wang, Y.L. (eds) (1989) *Fluorescence Microscopy of Living Cells in Culture, Parts A and B (Methods in Cell Biology, Volumes 29 and 30).* Academic Press, San Diego, CA.

Taylor, D.L., Nederlof, M., Lanni, F. and Waggoner, A.S. (1992) The new vision of light microscopy. *Am. Sci.* **80**: 322–335.

Taylor, D.L., Waggoner, A.S., Murphy, R.F., Lanni, F. and Birge, R.R. (eds) (1986) *Applications of Fluorescence in the Biomedical Sciences.* Alan R. Liss, New York.

Tsien, R.Y. (1994) Fluorescence imaging creates a window on the cell. *Chem. Eng. News* **72**: 34–36.

Wang, X.F. and Herman, B. (eds) (1996) *Fluorescence Imaging Spectroscopy and Microscopy,* Vol. 137 in Chemical Analysis: a Series of Monographs on Analytical Chemistry and Its Applications (J.D. Winefordner, ed.). John Wiley and Sons, New York.

Whery, E.L. (1981) *Modern Fluorescence Spectroscopy,* Vol. 1-4. Plenum Press, New York.

Wilson, T. (1990) *Confocal Microscopy.* Academic Press, San Diego, CA.

Zernicke, F. (1942) Phase contrast, a new method for the microscopic observation of transparent objects. *Physica* **9**: 686–693.

Zernicke, F. (1955) How I discovered phase contrast. *Science* **121**: 345–349.

Chapter 1 Fundamentals of fluorescence

General

Andrews, C.L. (1990) *Optics of the Electromagnetic Spectrum.* Prentice-Hall, Englewood Cliffs, NJ.

Dewey, T.G. (1991) *Biophysical and Biochemical Aspects of Fluorescence.* Plenum Press, New York.

Ditchburn, R.W. (1963) *Light.* Wiley Interscience, New York.

Gregory, S.S. (1977) *Fluorescence and Phosphorescence Spectroscopy: Physiochemical Principles and Practice.* Pergamon Press, Oxford.

Guilbault, G.G. (1973) *Practical Fluorescence – Theory, Methods and Techniques.* Marcel Dekker, New York.

Lakowicz, J.R. (1983) *Principles of Fluorescence Spectroscopy*. Plenum Press, New York.

Lakowicz, J.R. (ed.) (1991) *Topics in Fluorescence Spectroscopy: Techniques (Volume 1); Principles (Volume 2); Biochemical Applications (Volume 3)*. Plenum Press, New York.

Pesce, A.J., Rosen, C.-G. and Pasby, T.L. (1971) *Fluorescence Spectroscopy: An Introduction for Biology and Medicine*. Marcel Dekker, New York.

Rendell, D. (1987) *Fluorescence and Phosphorescence*. John Wiley and Sons, New York.

Quantum yield

Chen, R.F. (1965) Fluorescence quantum yield measurements: vitamin B_6 compounds. *Science* **150**: 1593–1595.

Schreiner, A.F. (1971) in *Fluorescence Spectroscopy: An Introduction for Biology and Medicine* (A.J. Pesce, C.-G. Rosen and T. Pasby, eds), p. 150. Marcel Dekker, New York.

Chen, R.F. (1974) Fluorescence lifetime reference standards for the range 0.189 to 115 nanoseconds. *Analytical Biochem.* **57**: 593–604.

Fluorescence lifetime

Isenberg, I. (1975) in *Biochemical Fluorescence: Concepts*, Vol. 1 (R.F. Chen and H. Edelhock, eds), pp. 43–77. Marcel Dekker, New York.

Ware, W.R. (1971) in *Creation and Detection of the Excited State*, Vol. 1, Part A (A.A. Lamola, ed.), p. 213. Marcel Dekker, New York.

Yguerabide, J. (1972) Nanosecond fluorescence spectroscopy of macromolecules. *Meth. Enzymol.* **26**: 498–578.

Polarization

Perrin, F.J. (1927) Fluoresece et introduction moléculaire par résonance. *C.R. Hebd. Seances Acad. Sci.* **184**: 1097–1100 .

Dale, R.E., Chen, L. and Brand, L. (1977) Rotational relaxation of the 'microviscosity' probe diphenylhexatriene in paraffin oil and egg lecithin vesicles. *J. Biol. Chem.* **252**: 7500–7510.

Heyn, M.P. (1979) Determination of lipid order parameters and rotational correlation times from fluorescence depolarization experiments. *FEBS Lett.* **108**: 359–364.

Gottlieb, Y.Y. and Wahl, P. (1963) Raman scattering study of the effect of pressure and temperature on the order–disorder phase transition in ammonium chloride. *J. Chem. Phys.* **60**: 849–854.

Shinitsky, M., Dianoux, A.C., Gilter, C. and Weber, G. (1971) Microviscosity and order in the hydrocarbon region of micelles and membranes determined with fluorescent probes. *Biochem.* **10**: 2106–2113

Wahl, P. and Weber, G. (1967) fluorescence depolarization of rabbit gamma globulin conjugates. *J. Mol. Biol.* **30**: 371–382.

Weber, G. (1952) Polarization of the fluorescence of macromolecules 2. Fluorescent conjugates of ovalbumin and bovine serum albumin. *Biophys. J.* **51**: 155–167.

Fluorescence resonance energy transfer (FRET)

Selvin, P.R. (1995) Fluorescence resonance energy transfer. *Meth. Enzymol.* **246**: 300–334.

Wu, P. and Brand, L. (1994) Resonance energy transfer: methods and applications (review). *Anal. Biochem.* **218**: 1–13.

Herman, B. (1989) Resonance energy transfer microscopy. In *Methods in Cell Biology* Vol. 30 (D.L. Taylor and Y.-L. Wang, eds), pp. 219–243. Academic Press, San Diego, CA.

Stryer, L. (1978) Fluorescence energy transfer as a spectroscopic ruler. *Ann. Rev. Biochem.* **47**: 819–846.

Dale, R.E. and Eisenger, J. (1975) in *Biochemical Fluorescence: Concepts*, Vol. 1 (R.F. Chen and H. Edelhock, eds), pp. 115–284. Marcel Dekker, New York.

Forster, T. (1959) Transfer mechanisms of electronic excitation. *Disc. Faraday Soc.* **27**: 7–17.

Chapter 2   Fluorescence microscopy

General

Periasamy, A. and Herman, B. (1994) Computerized microscopic vision in cell biology. *J. Comp. Assist. Micros.* **6**: 1–26.

Abramowitz, M. (1993) *Fluorescence Microscopy – The Essentials.* Olympus-America, New York.

Becker, E. (1989) *Fluorescence Microscopy.* Wild Leitz, Wetzlar.

Bradbury, S. and. Turner, G.L.E. (1967) *Historical Aspects of Microscopy.* W. Heffer and Sons, Cambridge.

Bradbury, S. (1967) *The Evolution of the Microscope.* Pergamon Press, Oxford.

Clark, G.L. (1961) *The Encyclopedia of Microscopy.* Reinhold, New York.

Clay, R.S. and Court, T.H. (1932) *The History of the Microscope.* Chas. Griffin and Co., London.

Hartley, W.J. (1964) *How to use a Microscope.* Doubleday, New York.

Herman, B. and Lemasters, J.J. (1993) *Optical Microscopy – Emerging Methods and Applications.* Academic Press, San Diego, CA.

Herman, B. (1998) *Fluorescence Microscopy*. Bios Scientific Publishers, Oxford.

Herman, B. and Jacobson, K. (1990) *Optical Microscopy for Biology*. Wiley/Liss, New York.

James, J. and Tanke, H.J. (1987) *Biomedical Light Microscopy*. Kluwer Academic, The Netherlands.

Kapitza, H.G. (1996) *Microscopy From the Very Beginning*. Carl Zeiss, Jena.

Lacey, A.J. (1989) *Light Microscopy in Biology*. IRL Press, Oxford.

Marmasse, C. (1980) *Microscopes and Their Uses*. Gordon and Breach, Oxon.

Mason, W.T. (1993) *Fluorescent and Luminescent Probes for Biological Activity*. Academic Press, London.

Mollring, F.K. (1981) *Microscopy from the Very Beginning*. Carl Zeiss, New York.

Patzelt, W.J. (1985) *Polarized Light Microscopy: Principles, Instruments, Applications*, 3rd Edn. E. Leitz, Wetzlar.

Payne, B.O. (1957) *Microscope Design and Construction*. Cooke, Troughton and Sims, York.

Pluta, M. (1989) *Advanced Light Microscopy*, Vols 1, 2 and 3. Elsevier, New York.

Reichman, J. (1994) *Glossary of Related Terminology with Page References to the Handbook of Optical Filters for Fluorescence Microscopy*. Chroma Technology Corp., Brattleboro, VT.

Reichman, J. (1994) *Handbook of Optical Filters for Fluorescence Microscopy*. Chroma Technology Corp., Brattleboro, VT.

Richardson, J.H. (1991) *Handbook for the Light Microscope: A User's Guide*. Noyes, New York.

Ross, K.F.A. (1967) *Phase Contrast and Interference Microscopy for Cell Biologists*. St Martin's Press, New York.

Rost, F.W.D. (1992) *Fluorescence Microscopy*, Vols 1 and 2. Cambridge University Press, Cambridge.

Rubbi, C.P. (1994) *Light Microscopy: Essential Data*. John Wiley and Sons, New York.

Shurcliffe, W.A. (1962) *Polarized Light*. Harvard Press, Cambridge, MA.

Shurcliffe, W.A. and Ballard, S.S. (1964) *Polarized Light*. Van Nostrand, Princeton, NJ.

Slavik, J. (1994) *Fluorescent Probes in Cellular and Molecular Biology*. CRC Press, Boca Raton, FL.

Slavik, J. (ed.) (1996) *Fluorescence Microscopy and Fluorescent Probes*. Plenum Press, New York.

Slayter, E.M. (1970) *Optical Methods in Biology*. Wiley Interscience, New York.

Slayter, E.M. and Slayter, H.S. (1992) *Light and Electron Microscopy*. Cambridge University Press, Cambridge.

Smith, R.F. (1994) *Microscopy and Photomicrography: A Working Manual*. CRC/Lewis Publishers, Boca Raton, FL.

Spencer, M. (1982) *Fundamentals of Light Microscopy*. Cambridge University Press, Cambridge.

Swindell, W. (1975) *Polarized Light*, Benchmark Papers in Optics. Halsted Press/Wiley, New York.

Tanke, H.J. (1995) *Fluorescence Microscopy (Microscope Handbooks, No. 35)*. Coronet Books, UK.

Taylor, D.L. and Wang, Y.L. (eds) (1989) *Fluorescence Microscopy of Living Cells in Culture, Parts A and B (Methods in Cell Biology, Volumes 29 and 30)*. Academic Press, San Diego, CA.

Taylor, D.L., Waggoner, A.S., Murphy, R.F., Lanni, F., and Birge, R.R. (eds) (1986) *Applications of Fluorescence in the Biomedical Sciences*. Alan R. Liss, New York.

Tsien, R.Y. (1994) Fluorescence imaging creates a window on the cell. *Chem. Eng. News* **72**: 34–36.

Wang, X.F. and Herman, B. (eds) (1996) *Fluorescence Imaging Spectroscopy and Microscopy*, Vol. 137 in Chemical Analysis: a Series of Monographs on Analytical Chemistry and Its Applications (J.D. Winefordner, ed.). John Wiley and Sons, New York.

Whery, E.L. (1981) *Modern Fluorescence Spectroscopy*, Vol. 1-4. Plenum Press, New York.

Optics

Andrews, C.L. (1960) *Optics of the Electromagnetic Spectrum*. Prentice-Hall Inc., Englewood Cliffs, NJ.

Born, M. and Wolf, E. (1967) *Principles of Optics*. Pergamon Press, Oxford.

Francon, M. (1966) *Diffraction Coherence in Optics*. Pergamon Press, Oxford.

Francon, M. (1966) *Optical Interferometry*. Academic Press, New York.

Hecht, E. (1987) *Optics*, 2nd Edn. Addison-Wesley, Reading, MA.

Jenkins, F.A. and White, H.E. (1976) *Fundamentals of Optics*, 4th Edn. McGraw-Hill, New York.

Swindell, W. (1975) *Polarized Light*, Benchmark Papers in Optics. Halsted Press/Wiley, New York.

Light sources

Ellis, G.W., Inoué, S. and Inoué, T. (1986) Computer-aided light microscopy. *Soc. Gen. Physiol. Ser.* **40**: 15–30.

Berns, M.W. (1974) *Biological Microirradiation, Classical and Laser Sources*. Prentice-Hall, Englewood Cliffs, NJ.

Inoué, S. and Hyde, W.L. (1957) Studies on depolarization of light at microscope lens surfaces. *J. Biophys. Biochem. Cytol.* **3**: 831–838.

Wavelength selection devices

Wang, X. and Lewis, E. N. (1996) in *Fluorescence Imaging Spectroscopy and Microscopy* (X.F. Wang and B. Herman, eds), pp. 125–156. Wiley-Interscience, New York.

Chatton, J.-Y. and Spring, K.R. (1993) Light sources and wavelength selection for widefield fluorescence microscopy. *MSA Bull.* **23**: 324–333.

Morris, H.R., Hoyt, C.C. and Treado, P.J. (1994) Imaging spectrometer for fluorescence and Raman microscopy: acousto-optic and liquid crystal tunable filters. *Appl. Spectrosc.* **48**: 857–866

Reichman, J. (1994) *Glossary of Related Terminology with Page References to the Handbook of Optical Filters for Fluorescence Microscopy*. Chroma Technology Corp., Brattleboro, VT.

Reichman, J. (1994) *Handbook of Optical Filters for Fluorescence Microscopy*. Chroma Technology Corp., Brattleboro, VT.

Objectives

Inoué, S. and Spring, K. (1997) *Video Microscopy: The Fundamentals.* Plenum Press, New York.

Keller, H.E. (1990) in *Handbook of Biological Confocal Microscopy*, (J. Pawley, ed.), pp. 77–86. Plenum Press, New York.

Inoué, S. and Oldenbourg, R. (1995) in *Handbook of Optics*, Vol. II, 2nd Edn (M. Bass, ed.), Chap. 17. McGraw-Hill, New York.

Detectors

Inoué, S. and Spring K. *Video Microscopy: The Fundamentals*. Plenum Press, New York.

Salmon, E.D., Walker, R.A. and Pryer, N.K. (1989) Video-enhanced differential interference contrast light microscopy. *Biotech.* **7**: 624–633.

Salmon, E.D. (1995) VE-DIC light-microscopy and the discovery of kinesin. *Trends Cell Biol.* **5**: 154–158.

Spring, K.R. and Lowy, R.J. (1989) Characteristics of low light level television cameras. *Meth. Cell Biol.* **29**: 269–289.

Tsay, T-T., Inman, R., Wray, B.E., Herman, B. and Jacobson, K. (1990) Characterization of low light level cameras for fluorescence microscopy. In *Optical Microscopy for Biology* (B. Herman and K. Jacobson, eds), pp. 219–234. Wiley-Liss, New York.

Chapter 3   Practical fluorescence microscopy

General

Abramowitz, M. (1993) *Fluorescence Microscopy – The Essentials.* Olympus-America, New York.

Becker, P.L. and Fay, F.S. (1987) Photobleaching of Fura-2 and its effect on the determination of calcium concentrations. *Am. J. Physiol.* **253**: C613–C618.

Hartley, W.J. (1964) *How to use a Microscope.* Doubleday, New York.

Kapitza, H.G. (1996) *Microscopy From the Very Beginning.* Carl Zeiss, Jena.

Mason, W.T. (1993) *Fluorescent and Luminescent Probes for Biological Activity.* Academic Press, London.

Moore, E.D.W., Becker, P.L., Fogarty, K.E., Williams, D.A. and Fay, F.S. (1990) Ca^{2+} imaging in single living cells: Theoretical and practical issues. *Cell Calcium* **11**: 157–179.

Richardson, J.H. (1991) *Handbook for the Light Microscope: A User's Guide.* Noyes, New York.

Smith, R.F. (1994) *Microscopy and Photomicrography: A Working Manual.* CRC/Lewis Publishers, Boca Raton, FL.

Taylor, D.L. and Wang, Y.L. (eds) (1989) *Fluorescence Microscopy of Living Cells in Culture, Parts A and B (Methods in Cell Biology, Volumes 29 and 30).* Academic Press, San Diego, CA.

Chapter 4   Applications of fluorescence microscopy

Immunofluorescence

Voss, E. Jr (1984) in *Fluorescein Hapten: An Immunologic Probe,* Uniscience Series. CRC Press, Boca Raton, FL.

Haugland, R.P. (1995) *Monoclonal Antibody Protocols (Methods in Molecular Biology, Volume 45)* (W.C. Davis, ed.), pp. 205–221. Humana Press, Totawa, NJ.

Hermanson, G.T. (1996) *Bioconjugate Techniques.* Academic Press, San Diego, CA.

Johnson, I.D., Ryan, D. and Haugland, R.P. (1993) *Methods in Nonradioactive Detection* (G.C. Howard, ed.), pp. 47–68. Appleton and Lange, Stamford, CT.

Hemmilä, I.A. (1991) *Applications of Fluorescence in Immunoassays.* John Wiley and Sons, New York.

Jameson, D.M. and Reinhart, G.D. (eds) (1989) *Fluorescent Biomolecules.* Plenum Press, New York.

Fluorescence in situ *hybridization (FISH)*

Kricka, L.J. (ed.) (1995) *Nonisotopic Probing, Blotting and Sequencing.* Academic Press, San Diego, CA.

Choi, Y.J. (1991) Detection of human papillomavirus DNA on routine Papanicolaou's smears by *in situ* hybridization with the use of biotinylated probes. *Am. J. Clin. Pathol.* **95**: 475–480.

Siadat-Pajouh, M., Periasamy, A., Ayscue, A.H., Moscicki, A.B., Palefsky, J.M., Walton, L., DeMars, L.R., Power, J.D., Herman, B. and Lockett, S.J. (1994) Detection of human papilloma virus type 16/18 DNA in cervicovaginal cells by fluorescence based *in situ* hybridization and automated image cytometry. *Cytometry* **15**: 245–257.

Siadat-Pajouh, M., Ayscue, A.H., Periasamy, A. and Herman, B. (1994) Introduction of a fast and sensitive fluorescent *in situ* method for single copy detection of human papillomavirus (HPV) genome. *J. Histochem.* **42**: 1503–1512.

Wilkinson, D.G. (1994) *In Situ Hybridization. A Practical Approach.* IRL Press, Oxford.

Fluorescence/differential interference contrast

Chatton, J.Y. and Spring, K.R. (1995) The sodium concentration of the lateral intercellular spaces of MDCK cells: a microspectrofluorimetric study. *J. Memb. Biol.* **144**: 11–19.

Harris, P.J., Chatton, J.Y., Tran, P.H., Bungay, P.M. and Spring, K.R. (1994) pH, morphology, and diffusion in lateral intercellular spaces of epithelial cell monolayers. *Am. J. Physiol.* **266**: C73–C80.

Foskett, J.K. (1988) Simultaneous Nomarski and fluorescence imaging during video microscopy of cells. *Am. J. Physiol.* **255**: C566–C571.

Spring, K.R. (1990) in *Optical Microscopy for Biology* (B. Herman and K. Jacobson, eds), pp. 513–522. Alan R. Liss, New York.

Automated fluorescence image cytometry

Lockett, S.J. and Herman, B. (1994) The automatic detection of clustered, fluorescent-stained nuclei by digital image based cytometry. *Cytometry* **17**: 1–12.

Lockett, S.J., Jacobson, K. and Herman, B. (1992) Quantification accuracy of an automated, fluorescence-based image cytometer. *Anal. Quat. Cytol. Histol.* **14**: 187–202.

Lockett, S.J., O'Rand, M., Wray, B.E., Rinehart, C., Kaufman, D., Herman, B. and Jacobson, K.A. (1991) Automated image-based cytometry: quantitation of DNA ploidy and detection of chlamydiae-infected cells. *Analyt. Quant. Cytol. Histol.* **13**: 27–44.

Lockett, S.J., Jacobson, K., O'Rand, M., Kaufman, D.G., Corcoran, M., Simonsen, M.G. and Herman, B. (1990) Automated image-based cytometry with fluorescent stained specimens. *Biotechniques* **10**: 514–519.

Wodnicki, P., Lockett, S., Pillai, M.R., Halabi, S., McKalip, A. and Herman, B. (1996) Automated image microscopy. In *Fluorescence*

Spectroscopy and Microscopy (B. Herman and X.F. Wang, eds), pp. 31–54. John Wiley and Sons, New York.

Fluorescence recovery after photobleaching (FRAP)

Zhang, F., Lee, G.M. and Jacobson, K. (1993) Protein lateral mobility as a reflection of membrane microstructure. *Bioessays* **15**: 579–588.

Zhang, F., Crise, B., Su, B., Hou, Y., Rose, J.K., Bothwell, A. and Jacobson, K. (1991) Lateral diffusion of membrane-spanning and glycosylphosphatidylinositol-linked proteins: toward establishing rules governing the lateral mobility of membrane proteins. *J. Cell Biol.* **115**: 75–84.

Jacobson, K., Zhang, F. and Tsay, T.T. (1991) Fluorescence recovery after photobleaching techniques to measure translational mobility in microscopic samples. *Scanning Micros.* **5**: 357–361.

Jacobson, K., Ishihara, A. and Inman, R. (1987) Lateral diffusion of proteins in membranes. *Ann. Rev. Physiol.* **49**: 163–175.

Webb, W.W., Barak, L.S., Tank, D.W. and Wu, E.S. (1981) Molecular mobility on the cell surface. *Biochem. Soc. Symp.* **46**: 191–205.

Gordon, G.W., Chazotte, B., Wang, X.F. and Herman, B. (1995) Analysis of simulated and experimental fluorescence recovery after photobleaching. Data for two diffusing components. *Biophys. J.* **68**: 766–778.

Total internal reflectance fluorescence microscopy (TIRF)

Tamm, L.K. (1993) in *Optical Microscopy: Emerging Methods and Applications* (B. Herman and J.J. Lemasters, eds), pp. 295–338. Academic Press, San Diego, CA.

Axelrod, D., Burghardt, T.P. and Thompson, N.L. (1984) Total internal reflection fluorescence. *Ann. Rev. Biophys. Bioeng.* **13**: 247–268.

Axelrod, D. (1981) Cell–substrate contacts illuminated by total internal reflection fluorescence. *J. Cell Biol.* **89**: 141–145.

Lanni, F., Waggoner, A.S. and Taylor, D.L. (1985) Structural organization of interphase 3T3 fibroblasts studied by total internal reflection fluorescence microscopy. *J. Cell Biol.* **100**: 1091–1102.

Tamm, L.K. (1988) Lateral diffusion and fluorescence microscope studies on a monoclonal antibody specifically bound to supported phospholipid bilayers. *Biochem.* **27**: 1450–1457.

Fluorescence resonance energy transfer microscopy (FRETM)

Herman, B. (1989) Resonance energy transfer microscopy. In *Methods in Cell Biology*, Vol. 30 (D.L. Taylor and Y.-L. Wang, eds), pp. 219–243. Academic Press, San Diego, CA.

Wang, X.F., Lemasters, J.J. and Herman, B. (1994) Plasma membrane architecture during hypoxic injury in rat hepatocytes measured by fluorescence quenching and resonance energy transfer. *J. Biomag.* **1**: 30–39.

Gadella, T.W., Jovin, T.M. and Clegg, R.M. (1994) Fluorescence lifetime imaging microscopy (FLIM): Spatial resolution of microstructures on the nanosecond time scale. *Biophys. Chem.* **48**: 221–239.

Jovin, T. M. and Arndt-Jovin, D. (1989) *Cell Structure and Function by Microspectrofluorometry.* Academic Press, San Diego, CA.

Wu, P. and Brand, L. (1994) Resonance energy transfer: methods and applications. *Anal. Biochem.* **218**: 1–13.

Uster, P.S. (1993) in *Methods in Enzymology: Membrane Fusion Techniques Part B* (N. Duzgunes, ed.), Vol. 221, pp. 239–246. Academic Press, San Diego, CA.

Clegg, R.M. (1996) in *Fluorescence Imaging Spectroscopy and Microscopy*, Chemical Analysis Series, Vol. 137 (X.F. Wang and B. Herman, eds), pp. 179–252. John Wiley and Sons, New York.

Digitized fluorescence polarization microscopy

Axelrod, D. (1989) in *Fluorescence Microscopy of Living Cells in Culture*, Methods in Cell Biology, Vol. 30 (D.-L. Taylor and Y.-L. Wang, eds), pp. 245–270. Academic Press, San Diego, CA.

Florine-Casteel, K. (1990) Phospholipid order in gel- and fluid-phase cell-size liposomes measured by digitized video fluorescence polarization microscopy. *Biophys. J.* **57**: 1199–1215.

Florine-Casteel, K., Lemasters, J.J. and Herman, B. (1991) Lipid order in hepatocyte plasma membrane blebs during ATP depletion measured by digitized video fluorescence polarization microscopy. *FASEB J.* **5**: 2078–2084.

Axelrod, D., Burghardt, T.P. and Thompson, N.L. (1984) Total internal reflection fluorescence. *Ann. Rev. Biophys. Bioeng.* **13**: 247–268.

Axelrod, D. (1981) Cell-substrate contacts illuminated by total internal reflection fluorescence. *J. Cell Biol.* **89**: 141–145.

Thompson, N.L., McConnell, H.M. and Burghardt, T.P. (1984) Order in supported phospholipid monolayers detected by the dichroism of fluorescence excited with polarized evanescent illumination. *Biophys. J.* **46**: 739–747.

Fluorescence lifetime imaging microscopy (FLIM)

Oida, T., Sato, Y. and Kusumi, A. (1993) Fluorescence lifetime imaging microscopy (flimscopy). Methodology, development and application to studies of endosome fusion in single cells. *Biophys. J.* **64**: 676–685.

Periasamy, A., Wang, X.F., Wodnicki, P., Gordon, G., Kwon, S., Diliberto, P.A. and Herman, B. (1995) High speed fluorescence microscopy: lifetime imaging in the biomedical sciences. *J. Micros. Soc. Am.* **1**:13–23.

Gadella, T.W., Jovin, T.M. and Clegg, R.M. (1994) Fluorescence lifetime imaging microscopy (FLIM): Spatial resolution of microstructures on the nanosecond time scale. *Biophys. Chem.* **48**: 221–239.

Wang, X.F., Periasamy, A., Coleman D. M. and Herman, B. (1992) Fluorescence lifetime imaging microscopy: Instrumentation and application. *Crit. Rev. Analyt. Chem.* **23**: 1–26.

Lakowicz, J.R. and Berndt, K.W. (1991) Lifetime-selective fluorescence imaging using an rf phase-sensitive camera. *Rev. Sci. Instrum.* **62**: 1727–1734.

Lakowicz, J. A., Szmacinski, H., Nowaczyk, K., Berndt, K.W. and Johnson, M.L. (1992) Fluorescence lifetime imaging. *Analy. Biochem.* **202**: 316–330.

Van de Ven, M. and Gratton, E. (1993) in *Optical Microscopy: Emerging Methods and Applications* (B. Herman and J.J. Lemasters, eds), pp. 373–402. Academic Press, San Diego, CA.

Fourier spectroscopy/spectral dispersion microscopy

Garini, Y., Katzir, N., Cabib, D., Buckwald, R.A., Soenksen, D.G. and Malik, Z. (1996) in *Fluorescence Imaging Spectroscopy and Microscopy*, Chemical Analysis Series, Vol. 137 (X.F. Wang and B. Herman, eds), pp. 87–124. John Wiley and Sons, New York.

Lewotsky, K. (1994) Hyperspectral imaging: evolution of imaging spectrometry. SPIE OE/Report, pp. 1–3. Bellvue, WA.

Ansersson-Engels, S., Johannson, J. and Svanberg, S. (1990) Multicolor fluorescence imaging systems for tissue diagnostics. *Proc. SPIE – Bioimag. Two-Dimens. Spectrosc.* **1205**: 179–189.

Malik, Z., Cabib, S., Buckwald, R.A., Garini, Y. and Soenksen, D. (1994) A novel spectral imaging system combining spectroscopy with imaging applications for biology. *Proc. SPIE – Opt. Imaging Tech. Biomed.* **2319**: 180–184.

Delayed luminescence microscopy

Periasamy, A., Siadat-Pajouh, M., Wodnicki, P., Wang, X.F. and Herman, B. (1995) Time-gated fluorescence microscopy for clinical imaging. *Micro. Anal.* 33–35.

Seveus, L., Vaisala, M., Syrjanen, S., Sandberg, M., Kuusisto, A., Harju, R., Salo, J., Hemmila, I., Kojola, H. and Soini, E. (1992) Time-resolved fluorescence imaging of europium chelate label in immunohistochemistry and *in situ* hybridization. *Cytometry* **13**: 329–338.

van Gijlswijk, R.P., Zijlmans, H.J., Wiegant, J., Bobrow, M.N., Erickson, T.J., Adler, K.E., Tanke, H.J. and Raap, A.K. (1997)

Fluorochrome-labeled tyramides: use in immunocytochemistry and fluorescence *in situ* hybridization. *J. Histochem. Cytochem.* **45**: 375–382.

Hennink, E.J., de Haas, R., Verwoerd, N.P. and Tanke, H.J. (1996) Evaluation of a time-resolved fluorescence microscope using a phosphorescent Pt-pophyrine model system. *Cytometry* **24**: 312–320.

Schauerte, J.A., Gafni, A. and Steel, D.G. (1996) Improved differentiation between luminescence decay components by use of time-resolved optical activity measurements and selective lifetime modulation. *Biophys. J.* **70**: 1996–2000.

Green fluorescent protein (GFP)

Chalfie, M., Tu, Y. and Prasher, D.C. (1994) Green fluorescent protein as a marker for gene expression. *Science* **263**: 802–805.

Heim, R. and Tsien, R.Y. (1996) Engineering green fluorescent protein for improved brightness, longer wavelengths and fluorescence resonance energy transfer. *Curr. Biol.* **6**: 178–182.

Westphal, M., Jungbluth, A., Heidecker, M., Muhlbauer, B., Heizer, C., Schwartz, J.M., Marriott, G. and Gerisch, G. (1997) Microfilament dynamics during cell movement and chemotaxis monitored using a GFP-actin fusion protein. *Curr. Biol.* **7**: 176–183.

Kimata, Y., Iwaki, M., Lim, C.R. and Kohno, K. (1997) A novel mutation which enhances the fluorescence of green fluorescent protein at high temperatures. *Biochem. Biophys. Res. Commun.* **232**: 69–73.

Siemering, K.R., Golbik, R., Sever, R. and Haseloff, J. (1996) Mutations that suppress the thermosensitivity of green fluorescent protein. *Curr. Biol.* **6**: 1653–1663.

Yang, T.T., Cheng, L. and Kain, S.R. (1996) Optimized codon usage and chromophore mutations provide enhanced sensitivity with the green fluorescent protein. *Nucl. Acids Res.* **24**: 4592–4593.

Ropp, J.D., Donahue, C.J., Wolfgang-Kimball, D., Hooley, J.J., Chin, J.Y., Cuthbertson, R.A. and Bauer, K.D. (1996) Aequorea green fluorescent protein: simultaneous analysis of wild-type and blue-fluorescing mutant by flow cytometry. *Cytometry* **24**: 284–288.

Ormo, M., Cubitt, A.B., Kallio, K., Gross, L.A., Tsien, R.Y. and Remington, S.J. (1996) Crystal structure of the Aequorea victoria green fluorescent protein. *Science* **273**: 1392–1395.

Plautz, J.D., Day, R.N., Dailey, G.M., Welsh, S.B., Hall, J.C., Halpain, S. and Kay, S.A. (1996) Green fluorescent protein and its derivatives as versatile markers for gene expression in living Drosophila melanogaster, plant and mammalian cells. *Gene* **173**: 83–87.

Cormack, B.P., Valdivia, R.H. and Falkow, S. (1996) FACS-optimized mutants of the green fluorescent protein (GFP). *Gene* **173**: 33–38.

Yang, T.T., Kain, S.R., Kitts, P., Kondepudi, A., Yang, M.M. and Youvan, D.C. (1996) Dual color microscopic imagery of cells expressing the green fluorescent protein and a red-shifted variant. *Gene* **173**: 19–23.

Mitra, R.D., Silva, C.M. and Youvan, D.C. (1996) Fluorescence resonance energy transfer between blue-emitting and red-shifted excitation derivatives of the green fluorescent protein. *Gene* **173**: 13–17.

De Giorgi, F., Brini, M., Bastianutto, C., Marsault, R., Montero, M., Pizzo, P., Rossi, R. and Rizzuto, R. (1996) Targeting aequorin and green fluorescent protein to intracellular organelles. *Gene* **173**: 113–117.

Chapter 5   Quantitative fluorescence microscopy

General

Adams, S.R., Harootunian, A.T. and Tsien, R.Y. (1991) Fluorescence ratio imaging of cyclic AMP in single cells. *Nature (London)* **394**: 694–697.

Brand, L. and Gohlke, J.R. (1972) Fluorescence probes for structure. *Ann. Rev. Biochem.* **41**:843–68.

Bastiaens, P.I. and Jovin, T.M. (1996) Microspectroscopic imaging tracks the intracellular processing of a signal transduction protein: fluorescent-labeled protein kinase C beta I. *Proc. Natl Acad. Sci.* **93**: 8407–8412.

Hiraoka, Y., Sedat, J.W. and Agard, D.A. (1987) The use of a charge-coupled device for quantitative optical microscopy of biological structures. *Science* **238**: 36–41.

Jericevic, Z., Wiese, B., Bryan, J. and Smith, L.C. (1989) Validation of an imaging system: steps to evaluate and validate a microscope imaging system for quantitative studies. *Meth. Cell. Biol.* **30**: 47–83.

Ludwig, M., Hensel, N.F. and Hartzman, R.J. (1992) Calibration of a resonance energy transfer imaging system. *Biophys. J.* **61**: 845–857.

Mathies, R.A. and Stryer, L. (1986) Single-molecule fluorescence detection: A feasibility study using phycoerythrin. In *Application of Fluorescence in the Biomedical Sciences* (D.L. Taylor, A.S. Waggoner, R.F. Murphy, F. Lanni and R.R. Birge, eds), pp. 129–140. Alan R. Liss, New York.

Moore, E.D.W., Becker, P.L., Fogarty, K.E., Williams, D.A. and Fay, F.S. (1990) Ca^{2+} imaging in single living cells: Theoretical and practical issues. *Cell Calcium* **11**: 157–179.

Rost, F.W.D. (1991) *Quantitative Fluorescence Microscopy*. Cambridge University Press, Cambridge.

Rost, F.W.D. (1992) *Fluorescence Microscopy*, Vols 1 and 2. Cambridge University Press, Cambridge.

Chapter 6   Digitized video microscopy

General

Castleman, K.R. (1979) *Digital Image Processing*. Prentice Hall, Upper Saddle River, NJ.

Cheng, P.C., Wu, W.L., Lin, T.H. and Wu, J.L. (1994) *Multidimensional Microscopy*. Springer Verlag, Heidelberg.

Gonzales, R.C. and Wintz, P. (1987) *Digital Image Processing*, 2nd Edn. Addison-Wesley, Reading, MA.

Herman, B. and Lemasters, J.J. (1993) *Optical Microscopy – Emerging Methods and Applications*. Academic Press, San Diego, CA.

Herman, B. (1998) *Fluorescence Microscopy*. Bios Scientific Publishers, Oxford.

Herman, B. and Jacobson, K. (1990) *Optical Microscopy for Biology*. Wiley/Liss, New York.

Inoué, S. and Spring, K. *Video Microscopy: The Fundamentals*. Plenum Press, New York.

James, J. and Tanke, H.J. (1987) *Biomedical Light Microscopy*. Kluwer Academic, The Netherlands.

Kotyk, A. and Slavick, J. (1989) *Intracellular pH and its Measurement*. CRC Press, Boca Raton, FL.

Mason, W.T. (1993) *Fluorescent and Luminescent Probes for Biological Activity*. Academic Press, London.

Nuccitelli, R. and Deamer, D.W. (1982) *Intracellular pH – Its Measurement, Regulation, and Utilization in Cellular Functions*, Kroc Foundation Series, Vol. 15. Alan R. Liss, New York.

Nuccitelli, R. (ed.) (1994) *A Practical Guide to the Study of Calcium in Living Cells (Methods in Cell Biology, Volume 40)*. Academic Press, New York.

Piller, H. (1997) *Microscope Photometry*. Springer-Verlag, Berlin.

Ploem, J.S. and Tanke, H.J. (1987) *Introduction to Fluorescence Microscopy*. Oxford University Press, Oxford.

Russ, J.C. (1994) *The Image Processing Handbook*, 2nd Edn. CRC Press, London.

Sharif, N.A. (1993) *Molecular Imaging in Neuroscience: A Practical Approach*. IRL Press, Oxford.

Shotten, D. (1993) *Electronic Light Microscopy*. Wiley-Liss, New York.

Shurcliffe, W.A. (1962) *Polarized Light*. Harvard Press, Cambridge, MA.

Shurcliffe, W.A. and Ballard, S.S. (1964) *Polarized Light*. Van Nostrand, Princeton, NJ.

Slavik, J. (1994) *Fluorescent Probes in Cellular and Molecular Biology*. CRC Press, Boca Raton, FL.

Slavik, J. (ed.) (1996) *Fluorescence Microscopy and Fluorescent Probes.* Plenum Press, New York.

Slayter, E.M. and Slayter, H.S. (1992) *Light and Electron Microscopy.* Cambridge University Press, Cambridge.

Tanke, H.J. (1995) *Fluorescence Microscopy (Microscope Handbooks, No. 35).* Coronet Books, Oxfordshire.

Taylor, D.L. and Wang, Y.L. (eds) (1989) *Fluorescence Microscopy of Living Cells in Culture, Parts A and B (Methods in Cell Biology, Volumes 29 and 30).* Academic Press, San Diego, CA.

Taylor, D.L., Waggoner, A.S., Murphy, R.F., Lanni, F. and Birge, R.R. (eds) (1986) *Applications of Fluorescence in the Biomedical Sciences.* Alan R. Liss, New York.

Tsien, R.Y. (1994) Fluorescence imaging creates a window on the cell. *Chem. Eng. News* **72**: 34–36.

Wang, X.F. and Herman, B. (eds) (1996) *Fluorescence Imaging Spectroscopy and Microscopy*, Vol. 137 in Chemical Analysis: a Series of Monographs on Analytical Chemistry and Its Applications (J.D. Winefordner, ed.). John Wiley and Sons, New York.

Imaging detectors

Bright, G.R. and Taylor, D.L. (1986) in *Applications of Fluorescence in the Biomedical Sciences*, pp. 257–288. Alan R. Liss, New York.

Bookman, R.J. (1990) in *Optical Microscopy for Biology* (B. Herman and K. Jacobson, eds), pp. 235–250. Wiley-Liss, New York.

Reynolds, G.T. and Taylor, D.L. (1980) Image intensification applied to light microscopy. *Bioscience* **30**: 586–592.

Spring, K.R. and Lowy, R.J. (1989) Characteristics of low-light television cameras. In *Methods in Cell Biology*, Vol. 29 (Y. Wang and D.L. Taylor, eds), pp. 269–282. Academic Press, San Diego, CA.

Spring, K.R. and Smith, P.D. (1987) Illumination and detection systems for quantitative fluorescence microscopy. *J. Micros.* **147**: 265–278.

Tsay, T.-T., Inman, R., Wray, B.E., Herman, B. and Jacobson, K. (1990) Characterization of low light level cameras for fluorescence microscopy. In *Optical Microscopy for Biology* (B. Herman and K. Jacobson, eds), pp. 219–234. Wiley-Liss, New York.

Image acquisition and digital image processing

Young, I.T. (1989) Image fidelity: characterizing the image transfer function. In *Methods in Cell Biology*, Vol. 30 (Y.-L. Wang and D.L. Taylor, eds), pp. 2–45. Academic Press, San Diego, CA.

Wick, R.A. (1985) *Photonics Spectra* **May**: 133.

Tsuchiya, Y., Inuzuka, E., Kurono, T. and Hosada, M. (1985) Interpolative centroiding in CCD-based image photon counting systems. *Adv. Electronics Electron Phys.* **64A**: 21–32.

Giloh, H. and Sedat, J.W. (1982) Fluorescence microscopy: Reduced photobleaching of rhodamine and fluorescein protein conjugates by *n*-propyl gallate. *Science* **217**: 1252–1255.

Arndt-Jovin, D.J., Robert-Nicoud, M., Kaufman, S.J. and Jovin, T.M. (1985) Fluorescence digital imaging microscopy in cell biology. *Science* **230**: 247–256.

Baxes, G.R. (1984) *Digital Image Processing: A Practical Primer.* Prentice Hall, Englewood Cliffs, NJ.

Inoué, T. (1993) in *Electronic Light Microscopy* (D. Shotton, ed.), pp. 95–104. Wiley-Liss, New York.

Ion (ratio) imaging

Dunn, K.W., Mayor, S., Myers, J.N. and Maxfield, F.R. (1994) Applications of ratio fluorescence microscopy in the study of cell physiology. *FASEB J.* **8**: 573–582.

Adams, S.R., Harootunian, A.T., Buechler, Y.J., Taylor, S.S. and Tsien, R.Y. (1991) Fluorescence ratio imaging of cyclic AMP in single cells. *Nature (London)* **349**: 694–697.

Bright, G.R., Fisher, G.W., Rogowska, J. and Taylor, D.L. (1989) Fluorescence ratio imaging microscopy. In *Methods in Cell Biology,* Vol. 30 (Y.-L. Wang and D.L. Taylor, eds), pp. 157–192. Academic Press, San Diego, CA.

Chew, C.S. and Ljungström, M. (1993) Measurement and manipulation of oscillations in cytoplasmic calcium. In *Optical Microscopy: Emerging Methods and Applications* (B. Herman and J.J. Lemasters, eds), pp. 133–176, Academic Press, San Diego, CA.

Grynkiewicz, G., Poenie, M. and Tsien, R.Y. (1985) A new generation of Ca^{2+} indicators with greatly improved fluorescence properties. *J. Biol. Chem.* **260**: 3440–3450.

Roe, M.W., Lemasters, J.J. and Herman, B. (1990) Assessment of Fura-2 for measurements of cytosolic free calcium. *Cell Calcium* **11**: 63–73.

Rooney, T.A., Sass, E.J. and Thomas, A.P. (1990) Agonist-induced cytosolic calcium oscillations originate from a specific locus in single hepatocytes. *J. Biol. Chem.* **265**: 10792–10796.

Cheng, H., Lederer, M.R., Lederer, W.J. and Cannell, M.B. (1996) Calcium sparks and $[Ca^{2+}]i$ waves in cardiac myocytes. *Am. J. Physiol.* **270**: C148–C159.

Messler, P., Harz, H. and Uhl, R. (1996) Instrumentation for multi-wavelengths excitation imaging. *J. Neurosci. Met.* **69**: 137–147.

Bornfleth, H., Aldinger, K., Hausman, M., Jauch, A. and Cremer, C. (1996) Comparative genomic hybridization imaging by the one-chip true-color CCD camera kappa CF 15 MC. *Cytometry* **24**: 1–13.

DeBernardi, M.A. and Brooker, G. (1996) Single cell Ca^{2+}/cAMP cross-talk monitored by simultaneous Ca^{2+}/cAMP fluorescence ratio imaging. *Proc. Natl Acad. Sci.* **93**: 4577–4582.

Miyawaki, A., Llopis, J., Heim, R., McCaffery, J.M., Adams, J.S., Ikura, M. and Tsien, R.Y. (1997) Fluorescent indicators for Ca^{+2} based on green fluorescent proteins and calmodulin. *Nature* **388**: 882–87.

Multiparameter digitized video microscopy (MDVM)

Morris, S.J. (1993) in *Optical Microscopy: Emerging Methods and Applications* (B. Herman and J.J. Lemasters, eds), pp. 177–212. Academic Press, San Diego, CA.

DeBiasio, R., Bright, G.R., Ernst, L.A., Waggoner, A.S. and Taylor, D.L. (1987) Five-parameter fluorescence imaging: wound healing of living Swiss 3T3 cells. *J. Cell. Biol.* **105**: 1613–1622.

Lemasters, J.J., Nieminen, A.-L., Gores, G.J., Dawson, T.L., Wray, B.E., Kawanishi, T., Florine-Casteel, K., Bond, J. and Herman, B. (1990) Multiparameter digitized video microscopy of toxic and hypoxic injury in single cells. In *Optical Microscopy for Biology* (B. Herman and K. Jacobson, eds), pp. 523–534. Wiley-Liss, New York.

Martinez-Zauilan, R., Martinez, G.M., Latanzio, F. and Gilles, R.J. (1991) Simultaneous measurement of intracellular pH and Ca^{2+} using the fluorescence of SNARF-1 and fura-2. *Am. J. Physiol.* **260**: C297–C307.

Chacon, E., Reece, J.M., Nieminen, A.-L., Zahrebelski, G., Herman, B. and Lemasters, J.J. (1994) Distribution of electrical potential, pH, free Ca^{2+} and volume inside cultured rabbit cardiac myocytes during chemical hypoxia: A multiparameter digitized confocal microscopic study. *Biophys. J.* **66**: 942–952.

Caged compound release/photoactivation

Kao, J.P.Y. and Adams, S.R. (1993) Photosensitive caged compounds: design, properties and biological applications. In *Optical Microscopy: Emerging Methods and Applications* (B. Herman and J.J. Lemasters, eds), pp. 27–86. Academic Press, San Diego, CA.

Rossi, F.M. and Kao, J.P.Y. (1997) Nmoc-DBHQ, a new caged molecule for modulating sarcoplasmic/endoplasmic reticulum Ca^{2+} ATPase activity with light flashes. *J. Biol. Chem.* **272**: 3266–3271.

Lipp, P., Huser, J., Pott, L. and Niggli, E. (1996) Subcellular properties of triggered Ca^{2+} waves in isolated citrate-loaded guinea-pig atrial myocytes characterized by ratiometric confocal microscopy. *J. Physiol.* **497**: 599–610.

Adams, S.R. and Tsien, R.Y. (1993) Controlling cell chemistry with caged compounds. *Ann. Rev. Physiol.* **55**: 755–784.

McCray, J.A. and Trentham, D.R. (1989) Properties and uses of photoreactive caged compounds. *Ann. Rev. Biophys. Biophys. Chem.* **18**: 239–270.

Chapter 7   Single and multiphoton microscopy

General

Cheng, P.C., Wu, W.L., Lin, T.H. and Wu, J.L. (1994) *Multidimensional Microscopy*. Springer Verlag, Heidelberg.

Gu, M. (1996) *Principles of Three-Dimensional Imaging in Confocal Microscopes*. World Scientific, Singapore.

Herman, B. and Lemasters, J.J. (1993) *Optical Microscopy – Emerging Methods and Applications*. Academic Press, San Diego, CA.

Herman, B. (1998) *Fluorescence Microscopy*. Bios Scientific Publishers, Oxford.

Herman, B. and Jacobson, K. (1990) *Optical Microscopy for Biology*. Wiley/Liss, New York.

Kriete, A. (1992) *Visualization in Biomedical Microscopies: 3-D Imaging and Computer Applications*. VCH Publishing, Weinheim.

Mason, W.T. (1993) *Fluorescent and Luminescent Probes for Biological Activity*. Academic Press, London.

Pawley, J. (1995) *Handbook of Biological Confocal Microscopy*, 2nd Edn. Plenum Press, New York.

Stevens, J.K., Mills, L.R. and Trogadis, J.E. (eds) (1994) *Three-Dimensional Confocal Microscopy: Volume Investigation of Biological Specimens (Cell Biology)*. Academic Press, San Diego, CA.

Taylor, D.L. and Wang, Y.L. (eds) (1989) *Fluorescence Microscopy of Living Cells in Culture, Parts A and B (Methods in Cell Biology, Volumes 29 and 30)*. Academic Press, San Diego, CA.

Wilson, T. (1990) *Confocal Microscopy*. Academic Press, San Diego, CA.

Confocal microscopy

Ancin, H., Roysam, B., Dufresne, T.E., Chestnut, M.M., Ridder, G.M., Szarowski, D.H. and Turner, J.N. (1996) Advances in automated 3-D image analyses of cell populations imaged by confocal microscopy. *Cytometry* **25**: 221–234.

Sako, Y., Sekihata, A., Yanagisawa, Y., Yamamoto, M., Shimada, Y., Ozaki, K. and Kusumi, A. (1997) Comparison of two-photon excitation laser scanning microscopy with UV-confocal laser scanning microscopy in three-dimensional calcium imaging using the fluorescence indicator Indo-1. *J. Micros.* **185**: 9–20.

Ojcius, D.M., Niedergang, F., Subtil, A., Hellio, R. and Dautry-Varsat, A. (1996) Immunology and the confocal microscope. *Res. Immunol.* **147**: 175–188.

Petroll, W.M., Jester, J.V. and Cavanagh, H.D. (1996) *In vivo* confocal imaging. *Int. Rev. Exp. Pathol.* **36**: 93–129.

Schild, D. (1996) Laser scanning microscopy and calcium imaging. *Cell Calcium* **19**: 281–296.

Konig, K., So, P.T., Mantulin, W.W., Tromberg, B.J. and Gratton, E. (1996) Two-photon excited lifetime imaging of autofluorescence in cells during UVA and NIR photostress. *J. Micros.* **183**: 197–204.

Lemasters, J.J., Chacon, E., Ohata, H., Harper, I.S., Nieminen, A.L., Tesfai, S.A. and Herman, B. (1995) Measurement of electrical potential, pH, and free calcium ion concentration in mitochondria of living cells by laser scanning confocal microscopy. *Mitochondrial Biogenesis Genetics Pt A (Series: Methods Enzymol.)* **260**: 428–444.

Wang, S.S. and Augustine, G.J. (1995) Confocal imaging and local photolysis of caged compounds: dual probes of synaptic function. *Neuron* **15**: 755–760.

Sanders, R., Draaijer, A., Gerritsen, H.C., Houpt, P.M. and Levine, Y.K. (1995) Quantitative pH imaging in cells using confocal fluorescence lifetime imaging microscopy. *Anal. Biochem.* **227**: 302–308.

Michael, C. and Hester, R.B. (1995) Confocal microscopy: a new horizon for cytology. *Diag. Cytopathol.* **12**: 49–50.

Nie, S., Chiu, D.T. and Zare, R.N. (1994) Probing individual molecules with confocal fluorescence microscopy. *Science* **266**: 1018–1021.

Multiphoton microscopy

Potter, S.M. (1996) Vital imaging: two photons are better than one. *Curr. Biol.* **6**: 1595–1598.

Potter, S.M., Wang, C.M., Garrity, P.A. and Fraser, S.E. (1996) Intravital imaging of green fluorescent protein using two-photon laser-scanning microscopy. *Gene* **173**: 25–31.

Brakenhoff, G.J., Squier, J., Norris, T., Bliton, A.C., Wade, M.H. and Athey, B. (1996) Real-time two-photon confocal microscopy using a femtosecond, amplified Ti:sapphire system. *J. Micros.* **181**: 253–268.

Piston, D.W., Masters, B.R. and Webb, W.W. (1995) Three-dimensionally resolved NAD(P)H cellular metabolic redox imaging of the *in situ* cornea with two-photon excitation laser scanning microscopy. *J. Micros.* **178**: 20–27.

Gryczynski, I., Gryczynski, Z. and Lakowicz, J.R. (1997) Two-photon excitation by the evanescent wave from total internal reflection. *Anal. Biochem.* **247**: 69–76.

Sako, Y., Sekihata, A., Yanagisawa, Y., Yamamoto, M., Shimada, Y., Ozaki, K. and Kusumi, A. (1997) Comparison of two-photon excitation laser scanning microscopy with UV-confocal laser scanning microscopy in three-dimensional calcium imaging using the fluorescence indicator Indo-1. *J. Micros.* **185**: 9–20.

Ber, K.M., So, P.T., Chen, Y., Mantulin, W.W. and Gratton, E. (1996) Scanning two-photon fluctuation correlation spectroscopy: particle counting measurements for detection of molecular aggregation. *Biophys. J.* **71**: 410–420.

Xu, C., Zipfel, W., Shear, J.B., Williams, R.M. and Webb, W.W. (1996) Multiphoton fluorescence excitation: new spectral windows for biological nonlinear microscopy. *Proc. Natl Acad. Sci.* **93**: 10763–10768.

Szmacinski, H., Gryczynski, I. and Lakowicz, J.R. (1996) Three-photon induced fluorescence of the calcium probe Indo-1. *Biophys. J.* **70**: 547–555.

Bennett, B.D., Jetton, T.L., Ying, G., Magnuson, M.A. and Piston, D.W. (1996) Quantitative subcellular imaging of glucose metabolism within intact pancreatic islets. *J. Biol. Chem.* **271**: 3647–3651.

Niswender, K.D., Blackman, S.M., Rohde, L., Magnuson, M.A. and Piston, D.W. (1995) Quantitative imaging of green fluorescent protein in cultured cells: comparison of microscopic techniques, use in fusion proteins and detection limits. *J. Micros.* **180**: 109–116.

Williams, R.M., Piston, D.W. and Webb, W.W. (1994) Two-photon molecular excitation provides intrinsic 3-dimensional resolution for laser-based microscopy and microphotochemistry. *FASEB J.* **8**: 804–813.

Three-dimensional wide-field microscopy

Carrington, W.A., Lynch, R.M., Moore, E.D., Isenberg, G., Fogarty, K.E. and Fay, F.S. (1995) Super-resolution three-dimensional images of fluorescence in cells with minimal light exposure. *Science* **268**: 1483–1487.

Han, K.F., Gubbens, A.J., Sedat, J.W. and Agard, D.A. (1996) Optimal strategies for imaging thick biological specimens: exit wavefront reconstruction and energy-filtered imaging. *J. Micros.* **183**: 124–132.

Chen, H., Hughes, D.D., Chan, T.A., Sedat, J.W. and Agard, D.A. (1996) IVE (image visualization environment): a software platform for all three-dimensional microscopy applications. *J. Struct. Biol.* **116**: 56–60.

Scalettar, B.A., Swedlow, J.R., Sedat, J.W. and Agard, D.A. (1996) Dispersion, aberration and deconvolution in multi-wavelength fluorescence images. *J. Micros.* **182**: 50–60.

Isenberg, G., Etter, E.F., Wendt-Gallitelli, M.F., Schiefer, A., Carrington, W.A., Tuft, R.A. and Fay, F.S. (1996) Intrasarcomere $[Ca^{2+}]$ gradients in ventricular myocytes revealed by high speed digital imaging microscopy. *Proc. Natl Acad. Sci.* **93**: 5413–5418.

Loew, L.M., Tuft, R.A., Carrington, W. and Fay, F.S. (1993) Imaging in five dimensions: time-dependent membrane potentials in individual mitochondria. *Biophys. J.* **65**: 2396–2407.

Chapter 8 Fluorescence photomicroscopy

General

Castleman, K.R. (1979) *Digital Image Processing*. Prentice Hall, Upper Saddle River, NJ.

Gonzales, R.C. and Wintz, P. (1987) *Digital Image Processing*, 2nd Edn. Addison-Wesley, Reading, MA.

Loveland, R.P. (1981) *Photomicrography: A Comprehensive Treatise*, 2nd Edn, Vols 1 and 2. Wiley, New York.

Piller, H. (1977) *Microscope Photometry*. Springer-Verlag, Berlin.

Richardson, J.H. (1991) *Handbook for the Light Microscope: A User's Guide*. Noyes, New York.

Shotten, D. (1993) *Electronic Light Microscopy*. Wiley-Liss, New York.

Smith, R.F. (1994) *Microscopy and Photomicrography: A Working Manual*. CRC/Lewis Publishers, Boca Raton, FL.

Appendix G

University/educational WWW sites for microscopy and microanalysis

---

USA

A–H

Microscopy and Imaging Resources on the WWW Center for Toxicology, SW Environmental Health Sciences Center, University of Arizona
> http://www.pharm.ArizoNA.EDU/centers/tox_center/swehsc/ exp_path/m-i_onw3.ht.

Computer Networking Laboratory for Microscopy Education, University of Arizona, Tucson
> http://aluminum.sem.arizoNA.edu:8001/

Center for Solid State Science at Arizona State University
> http://www.asu.edu/clas/csss

National Center for Microscopy and Imaging Research (NCMIR) at University of California, San Diego
> http://www-ncmir.ucsd.edu

Electron Microscopy and Image Processing Lab, University of Chicago, Chicago, Illinois
> http://befvax.uchicago.edu:8000/~befty

Hausfeld–Kluener Microscopy Facility, at the University of Cincinnati, Department of Surgery
> http://www.med.uc.edu/htdocs/medicine/departme/hausfeld/kluener.htm

Electron Microscopy/Imaging Center, Colorado State University
> http://www.vetmed.colostate.edu/anatomy/emic/homepage.html

Center for BioMedical Imaging Technology, University of Connecticut
> http://panda.uchc.edu/htbit/

ICBR EM Core Laboratory at the University of Florida
> http://www.biotech.ufl.edu/~emcl/

Biological Electron Microscopy Facility at the University of Hawaii
http://www.pbrc.hawaii.edu/bemf/

I–M

Microscope Imaging Laboratory, University of Illinois, College of Veterinary Medicine at Urbana
http://www.cvm.uiuc.edu/announcements/MicSoc/MicImagLab.html
Center for Microanalysis of Materials, Fredrick Seitz Materials Research Lab, University of Illinois at Urbana
http://mrlxp2.mrl.uiuc.edu:80/~cmm/
Beckman Institute Visualization Facility, University of Illinois at Urbana
http://delphi.beckman.uiuc.edu/
Electron Microscope Facility, University of Illinois at Champagne Urbana
http://tigger.uic.edu/~gadget/RRC/emf.html
Central Microscopy Research Lab, University of Iowa
http://www.uiowa.edu/~cemrf/
Center for Microanalysis, University of Maryland at College Park, Maryland
http://www.glue.umd.edu/~junhui/lab/Welcome.html
Electron Microbeam Analysis Laboratory, University of Michigan, Ann Arbor
http://www.engin.umich.edu/lab/emal/
Characterization Facility, University of Minnesota
http://charfacnu.cie.umn.edu
Scanned Tip and Electron Image Lab, University of Missouri at St. Louis
http://newton.umsl.edu/stei_lab/stei_lab.html
Electron Beam Analytical Facility, University of Missouri
http://www.missouri.edu/~geosclmr/ebaf.html

N–Z

Bio-Imaging Center at Nassau Community College, Garden City, New York
http://www.sunynassau.edu/webpages/biology/becks.htm
Central Facility for Electron Microscopy at the University of Nebraska
http://www.unl.edu/tvoiles/
TEM Laboratory, University of New Mexico
http://tem.unm.edu
Advanced Microscopy and Imaging Lab, State University of New York, Buffalo
http://corn.eng.buffalo.edu/index.html
Pathology Department, Health Science Center, State University of New York, Syracuse
http://WWW.HSCSYR.EDU/~PATHENVI/
Stony Brook Microscopy Home Page, State University of New York, Stony Brook
http://xray1.physics.sunysb.edu/sbmic/index.html

University Microscopy Imaging Center, Stony Brook, New York
http://www.umic.sunysb.edu
Department of Cell Biology and Anatomy, Digitized Microscopy Facility,
University of North Carolina at Chapel Hill
http://www-cellbio.med.unc.edu/brochure/video.htm
NCSU SPM Labs Analytical Instrumentation Facility, NCSU, Raleigh,
North Carolina
http://spm.aif.ncsu.edu
HREM and Surface Structure Facility at Northwestern
http://risc1.numis.nwu.edu/internet/lab.html
Oklahoma Electron Microscopy Lab at University of Oklahoma
http://www.uoknor.edu/electron/
Electron Microprobe Facility, University of Oregon
http://darkwing.uoregon.edu:80/~mshaf/epmahome/
Structural Biology Imaging Center, University of Pittsburgh
http://sbic6.cbio.pitt.edu/sbic.htm
Purdue University Cytometry Laboratories, Confocal and Image
Analysis, Purdue University
http://www.cyto.purdue.edu
Purdue University Nanoscale Physics, West Lafayette, Indiana
http://www.physics.purdue.edu/nanophys
Purdue University Cryo-Electron Microscopy and Image Analysis
Laboratory
http://bilbo.bio.purdue.edu/~baker
Fermin Laboratory at Tulane University, Department of Pathology
http://www1.omi.tulane.edu/departments/pathology/fermin/cdftop.html
Biological Microscopy and Image Reconstruction Resource, Wadsworth
Center in Albany, New York
http://www.wadsworth.org/spider_doc/bmirr/cover.html
Integrated Microscopy Resource, University of Wisconsin, Madison
http://www.bocklabs.wisc.edu/imr.html
Center for Cell Imaging, Yale University School of Medicine, New
Haven, Connecticut
http://info.med.yale.edu/cellimg

North and South America

Canada

3D Laser Scanning Confocal Microscopy, Canada
http://www.cs.ubc.ca/spider/ladic/confocal.html

Casino Home Page (Monte Carlo Simulations) at the University of Sherbrooke, Canada

    http://www.gme.usherb.ca/casino

Medicine/Dentistry Electron Microscope Unit at the University of Alberta

    http://www.ualberta.ca/~mingchen/

Electron Microscopy Unit at the University of New Brunswick

    http://www.asg.unb.ca/emunit/emu.html

In Vitro Plus, Royal Victoria Hospital, McGill University, Montreal, Quebec

    http://libra.rvh.mcgill.ca

Europe

Belgium

Center for Electron Microscopy for Materials Research, University of Antwerp, Belgium

    http://zorro.ruca.ua.ac.be/%7EEMAT/

Finland

The Institute of Electron Optics, University of Oulu, Finland

    http://koivu.oulu.fi/~eolwww/welcome.html

University of Turku, Laboratory of Electron Microscopy, Finland

    http://www.utu.fi/med/tdk/em/index.html

France

Laboratoire de Mineralogie-Cristallographie, University of Paris VI-VII, Paris, France

    http://www.lmcp.jussieu.fr/index_en.html

Germany

Fritz-Haber-Institut der Max-Planck-Gesellschaft, Germany

    http://borg.rz-berlin.mpg.de/em/

Mikroskopische Arbeitsgemeinschaft Stuttgart eV, Microscopy Society of Stuttgart, Germany

    http://www.uni-hohenheim.de/~schwarzu/index.html

Electron Holography Project, University of Tuebingen, Germany

    http://gandalf.iap.physik.uni-tuebingen.de/holo/index.html

Netherlands

Electron Microscopy Group, University of Groningen, The Netherlands
http://rugbe2.chem.rug.nl/
Particle Optics Research Group, Technical University of Delft, The
Netherlands
http://wwwdo.tn.tudelft.nl/
National Centre for HREM, Technical University of Delft, The Netherlands
http://dutsm43.tudelft.nl/

Sweden

Electron Microscopy Unit, University Hospital, Lund, Sweden
http://www.emu.lu.se/

Switzerland

Centre Interdepartmental de Microscopie Electronique, Ecole
Polytechnique Federale de Lausanne, Switzerland
http://cimewww.epfl.ch/
The Laboratory for Electron Microscopy I, Department of Biology, Swiss
Federal Institute of Technology (ETH), Zurich, Switzerland
http://www.em.biol.ethz.ch/
Science Photography Lab, University of Basel, Switzerland
http://foto.chemie.unibas.ch/

UK

Department of Crystallography, Birbeck College, UK
http://www.cryst.bbk.ac.uk/
Microstructural Physics, University of Bristol, UK
http://www.phy.bris.ac.uk/research/microstructure/home.html
NWSTEM Facility, Liverpool University, UK
http://vgstem.mat.liv.ac.uk/nwstem.html
Royal Microscopy Society
http:www.blacksci.co.uk/products/journals/prms.html

Australia, Asia and the South Pacific

Australia

Analytical EM Facility, Queensland University of Technology (QUT),
Brisbane, Australia
http://www.sci.qut.edu.au/aemf/

Virus Data Base of Bioinformatics Facility at the Australian National University, Canberra, Australia

http://life.anu.edu.au/viruses/welcome.html

Electron Microscopy Facility, University of Tasmania, Hobart, Australia

http://lab.csl.utas.edu.au/facs/EMS.html

Diffraction Physics Research Laboratory at the School of Physics, University of Melbourne, Australia

http://www.ph.unimelb.edu.au/~djd/diff-home.html

Centre for Microscopy and Microanalysis, University of Queensland, Australia

http://www.uq.oz.au/nanoworld/nanohome.html

Electron Microscopy Unit, University of Sydney, Australia

http://www.usyd.edu.au/su/emu

Centre for Microscopy and Microanalysis, University of Western Australia, Nedlands, WA, Australia

http://cmm.uwa.edu.au/

Japan

Sakurai Laboratory, The IMR Atom Probe Group, Tohoku University, Japan

http://apfim.imr.tohoku.ac.jp/index.html

Wakabayashi Laboratory, University of Tokyo, Japan

http://tkyemghp.phys.s.u-tokyo.ac.jp/index.html

Middle East, Africa and other parts of the world

Middle East

Electron Probe Laboratory, Hebrew University of Jerusalem, Institute of Earth Sciences, Israel

http://earth.es.huji.ac.il/machon/e-prob/prob.html

Appendix H

Commercial and non-profit WWW sites for microscopy and microanalysis

A

Advanced Surface Microscopy, Inc.
  http://www.a1.com/asm/
Analytical Light Microscopy Services
  mailto: microls@aol.com
Applied Precision
  http://www.api.com/
ASTON Metallurgical Services Co., Inc.
  www.astonmet.com
AutoQuant
  http://aqi3.aqi.com

B

Bioptechs, Inc.
  http://www.bioptechs.com
BioRad Laboratories
  http://www.biorad.com/

C

Chroma
  http://www.chroma.com
Compix
  http://compix.com/

D

Dage MTI, Inc.
  http://www.dagemti.com/
Delaware Diamond Knives, Inc.
  http://www.ddk.com/
Diatome US
  http://www.emsdiasum.com/diatome
Digital Instruments, Inc.
  http://www.di.com

E

Edwards High Vacuum International
  http://www.edwards.boc.com
Electron Microscopy Sciences
  http://www.emsdiasum.com
Electroscan Corporation
  http://www.electroscan.com
EMiSPEC Systems, Inc.
  http://WWW.emispec.COM
Charles Evans & Associates
  http://www.cea.com/home.htm
Evex Analytical
  http://www.evex.com

F

Fisons Instruments
  http://www.fisonssurf.co.uk

Ernest F. Fullam, Inc.
  http://www.fullam.com/

G

Gatan, Inc.
  http://www.gatan.com
Geller Microanalytical Laboratory
  http://www.gellermicro.com
Goldmark Biologicals
  http://members.aol.com/goldmarker
GW Electronics, Inc.
  http://www.gwelectronics.com

H

Hamamatsu Photonic Systems
  http://www.hpk.co.jp/
NSA/Hitachi Scientific Instruments
  http://www.nissei.com
Hitachi Instruments, Inc.
  http://www.hii.hitachi.com
Hitachi Instruments, Inc., Japan
  http://www.hitachi.co.jp/Div/keisokuki

I

IBM Analytical Services
  http://www.chips.ibm.com/services/asg
ICMAS, Inc.
  http://www.usit.net/icmas/
International Equipment Trading Ltd
  http://www.ietltd.com
Intracellular Imaging
  http://intracellular.com/

J

JEOL USA, Inc.
  http://www.jeol.com

L

LEO
  http://www.mwrn.com/leo/
Leica Instruments
  http://www.Leica.com
Lomo Microscopes
  http://www.comet.net/gek

M

Materials Analytical Services
  http://www.mwrn.com/mas/
McCrone Group
  http://www.mccrone.com/
McCrone Research Institute
  http://www.mcri.org
Meridian Instruments
  http://mwrn.ms.wwa.com/page/meridian/meridian.html
MicroBrightField, Inc.
  http://www.microbrightfield.com
Microscopy & Analysis
  http://www.microrgc.demon.co.uk
Microscopy Today
  http://www.microscopy-today.com
Micro Star Diamond Knives
  http://www.microstartech.com/
MicroTherm, LLC
  http:// www.microthermllc.com
Molecular Imaging
  http://www.molec.com/

N

NanoProbes
http://www.nanoprobes.com
Nikon, Inc.
http://www.nikonusa.com/
Noran Instruments
http://www.noran.com
NT-MDT Co. (Molecular Devices and Tools for NanoTechnology)
http://www.ntmdt.ru

O

Olympus Corporation
http://www.ipg.olympusamerica.com/
Omega Optical
http://www.sover.net/~omega/
Optimas Corporation
http://www.optimas.com
Optronics Engineering
http://www.optronics.com
Optiscan P/L
http://www.optiscan.com.au
Oxford Instruments
http://www.oxinst.com/ri/index.html

P

Park Scientific Instruments
http://www.park.com
Philips Electronic instruments
http://WWW.peo.philips.com
Pixera Polysciences, Inc.
http://www.polysciences.com
PhotoMetrics, Inc.
http://www.wavenet.com/photomet/
Princeton Gamma-Tech, Inc.
http://www.pgt.com

Princeton Instruments
   http:www.prinst.com/
ProSciTech (formerly Probing and Structure)
   http://www.proscitech.com.au

R

Research Diagnostics, Inc.
   http://www.researchd.com/index.htm
Ross Electron Optics Consulting and Services
   http://www.missouri.edu/~geosclmr/lou.html

S

SEM-EDS Laboratory Metals d.o.o. Slovenia
   http://www2.arnes.si/guest/sgszmera1/index.html
Scanalytics
   http://www.scanalytics.com
Scientific Instrumentation Services, Inc.
   http://www.io.com/~ablue/
Scott Scientific Supplies
   http://www.scottscientific.com
SciTrak Buyers Guide
   http://www.scitrak.com
SD Spectral Diagnostics Inc.
   http://www.spectraldiagnostics.com/
Signal Analytics Corporation
   http://guide.nature.com/company/signalanalyticscorp
SIS (Soft-Imaging Software) GmbH
   http://www.soft-imaging-web.de
Small World
   http://members.aol.com/smworld100/index.htm
Spectra Services, Inc.
   http://www.frontiernet.net/~mspecht/
SPI Supplies
   http://www.2spi.com
South Bay Technology
   http://www.southbaytech.com
Synoptics
   http://www.synoptics.co.uk

T

Ted Pella, Inc.
  http://www.tedpella.com
TopoMetrix Inc.
  http://www.topometrix.com
Total Resolution
  http://www.TotalResolution.com
Tympel Engineering Office for Medical and Technical Image Processing
  http://home.t-online.de/home/One.Magic/homepadt.htm

U

Universal Imaging
  http://www.image1.com/

V

VayTek
  http://www.vaytek.com/
Virtual Laboratories
  http://www.Rt66.com/~virtlabs/
Vital Imaging
  http://www.vitalimages.com/

X

XEI Scientific
  http://www.msa.microscopy.com/SM/XEI/XEIHomePage.html
Xillix
  http://www.xillix.com/

Z

Zeiss Microscopy Division (USA)
  http://www.zeiss.com/micro/
Zeiss Microscopy Division (Germany)
  http://www.zeiss.de/mi/mi_e.html

Numeric

4 Pi Analysis
  http://www.4pi.com/

Appendix I: Glossary

Absorption: Interaction of light with matter, resulting in decreased intensity across entire spectrum or loss of intensity from a portion of the spectrum.

Achromat: Microscope objective corrected for axial chromatic aberration.

Acousto-optical tunable filter (AOTF): Radio-frequency is used to set up vibrations in a particular crystal material, generating a bulk transmission diffraction grating. By varying the radio frequency, it is possible to rapidly diffract out any desired wavelength of light.

A/D: Analogue-to-digital converter.

Airy disc: Three-dimensional diffraction pattern of object formed by each imaging point in the specimen. The overlap of neighbouring Airy discs determines the resolution of the microscope.

Angle of incidence: Angle between optical axis of light at the surface of a filter and the axis normal to this surface.

Angstrom (Å): An angstrom is 10^{-10} meters.

Anisotropy: The difference in a physical property for the same material measured along different directions. In microscopy, this usually refers to the differential transmission of light through a specimen.

Aperture diaphragm: Adjustable diaphragm in illumination path that regulates amount of excitation intensity (numerical aperture of excitation light).

Apochromat: Objective corrected for spherical and chromatic aberration.

Autofluorescence: Fluorescence from objects in a microscope sample other than from fluorophores.

Back focal plane: Focal plane located on the side of the objective away from the object.

Background: Fluorescence produced by anything in a microscopic sample other than from the desired fluorescent reporter molecules.

Background subtraction: Used in video image processing to subtract away persistent noise in an image.

Band-pass: Filter that passes light of a certain restricted range of wavelengths. Named by its centre wavelength and bandwidth.

Bandwidth (FWHM): Full width half maximal transmission. Separation between cut-on and cut-off wavelengths at 50% transmission. A

580/40 nm filter transmits maximally at 580 nm and transmits at 50% at both 560 and 600 nm.

Barrier (emission) filter: Filter that attenuates excitation light after passing through an excitation filter yet transmits emitted fluorescence.

Charge coupled device (CCD): A camera used for digital imaging. Contains a light-sensitive silicon chip. Light falling on the chip creates an electrical charge at a specific location.

Chromatic aberration: Inaccurate focusing of red, blue and green light either along or at right angles to the optical axis. Axial chromatic aberration results in the red image being focused further along the optical axis than the green image. Lateral chromatic aberration results in a slightly bluer image.

Coma: A lens aberration where off-axis beams do not form a single focused spot but a comet-shaped pattern.

Condenser: Optical component which regulates the angle and placement of light interacting with the specimen. It is placed between the specimen and the light source.

Conjugate focal planes: Planes that are in focus with each other. The object in the first focal plane forms an image in the second focal plane, which can then act as a object for the next focal plane, and so on.

Contrast: Visibility of an object against its background.

Critical illumination: Illumination optics in which the image of the light source is focused on the specimen plane.

Cross talk: Minimum attenuation level (over a specified wavelength range) of two filters put together in a series.

Dark-field: A type of diascopic illumination where the specimen is excited by light entering the condenser at oblique angles so as not to enter the objective directly.

Depth of field: Vertical distance between the closest and farthest objects that are all in focus in a sample. The depth of field varies as a function of the NA, wavelength of light and focal length of the lens.

Depth of focus: Vertical distance between a lens and the detector image plane for which the image formed by the lens is in focus.

Diascopic illumination: Illumination of specimen using light transmitted through the specimen and focused with a condenser.

Dichromatic mirror: Mirror in filter cube that reflects light below a certain wavelength (usually excitation light) but transmits light above a certain wavelength (usually emission light).

Dynamic range: Ratio of maximum to minimum signal levels in the absence of distortion.

Edge filter: Filter with very sharp cut-on or cut-off. Can be short-pass or long-pass filter.

Epi-illumination: Specimen is illuminated using light reflected on to it (usually from above).

Excitation filter: Filter that transmits only wavelengths of light required for excitation of a specific fluorophore.

Extinction coefficient: A measure of the amount of light that can be absorbed by a fluorophore.

Field diaphragm: A variable diaphragm located in the illumination pathway.

Filter glass (absorption glass): Glass which absorbs certain wavelengths of light and transmits other wavelengths of light.

Fluorescence: Property of certain molecules to absorb energy in the form of light and then release this energy at a longer wavelength than the wavelength of absorption (i.e. at a lower energy level).

Focal length: Distance between optical centre of a lens and the point at which it focuses light coming from infinity.

Focus: The ability of a lens to converge light rays to a single point.

Foot candle (fc): Amount of illumination from 1 international candle (candela) falling on a 1 ft^2 surface at a distance of 1 ft. The human eye can detect light levels as low as 10^{-4}–10^{-5} fc.

Frequency: The number of waves or signals that pass a certain point in a given amount of time.

Front surface: Side of a beam splitter which faces the excitation light.

Grey level: The brightness of pixels in a digitized video image. For an 8-bit signal, this ranges from 0 (black) to 225 (white).

Heat filter: Absorption glass filter that attenuates infrared radiation, but transmits light in the visible wavelength range.

Histogram: Graph showing the number of pixels displaying each (range of) grey level value(s).

Image processing: Various mathematical procedures to improve the signal-to-noise and contrast, and to obtain quantitative intensity data from images.

Immersion medium: Material placed between the uppermost surface of a sample and the objective.

Indirect fluorescence: Fluorescence emitted by fluorophores which are not an endogenous part of the specimen. Usually introduced into a specimen as a stain or probe.

Infinity (∞): Imaginary point far away from light source. Optically, infinity is operationally defined as a distance greater than 10 times the focal length of a lens.

Infinity-corrected optics: A two-lens optical design where the objective forms an image of an object at infinity which is then refocused by a second lens (Telen lens). Infinity-corrected optics allow great versatility in the design of the optical path, as addition or removal of optical components has no effect on the image.

Intensity: How bright an object is. Intensity is proportional to amplitude squared.

Interference: Interaction of one wave with another wave. Can be constructive or destructive.

Isotropic: Absorption and emission of light by the specimen is the same in all directions.

Köhler illumination: Illumination optics resulting in the image of the light source being out of focus at the specimen plane. Provides homogeneous illumination of the specimen.

Long working distance: Defines the amount of space between the optical components (condenser, objectives) and sample.

Long-pass filter: Filter that allows light greater than a certain wavelength to pass through the filter while preventing light below certain wavelengths from passing through the filter.

Magnification: Relationship between the size of an image and the size of the original object.

Micron (μm): A micron is 10^{-6} meters.

Nanometer (μm): A nanometer is 10^{-9} meters (= 10 Å).

Near infrared (NIR): Portion of electromagnetic spectum from 750 to 2500 nm.

Neutral density filter: Filter that attenuates the intensity of light independent of wavelength.

Numerical aperture (NA): Relationship between the objective light collection angle (α) and the refractive index (η) of the medium between the objective and specimen (NA = $\eta\sin\theta$).

Optical density (OD): Logarithmic unit of transmission. OD = $-\log T$ (transmission). A change of OD from 1 to 2 represents a 10-fold increase in absorption.

Optical memory disk recorder (OMDR): A video disk recorder employing a laser beam to write and read video frames on to/from storage medium. Operates at video frame rates.

Optical path: The physical distance or thickness light travels through a material times the refractive index of the material.

Photobleaching (fading): Photochemical reaction of fluorophore, light and oxygen that causes the intensity of the fluorescence emission to decrease with time.

Photometry: Measurement of light.

Photomultiplier tube (PMT): Converts incoming photons of light into electrons and then amplifies the number of electrons.

Photon: A packet of light energy.

Pixel: A picture (pix) element (el). Points of information used to map images. Pixels exist in arrays with specific x, y coordinates.

Polarization: Restriction of the orientation of the vibration of electromagnetic waves of light.

Polarizer: Material which can absorb or reflect light in all but one direction of vibration.

Polarizing filter: Filter that generates polarized light.

Polychroic: Dichroic beam splitter with multiple reflection bands and transmission regions.

Primary fluorescence: Fluorescence originating from the specimen itself.

Quantum efficiency: Measure of how efficiently a fluorophore converts absorbed energy into fluorescence.

Quenching: Any process that decreases the quantum efficiency of a fluorophore.

Raster scan: Process by which video cameras output information. Uses interlace format, where alternate raster lines are scanned by successive vertical sweeps.

Reflection: Light bouncing off an object in either a random or organized fashion.

Refraction: Bending of light as it passes through a material.

Refractive index (η): Relationship between velocity of light in a material of interest and the velocity of light in air or vacuum. The slower the velocity of light through a material, the higher the refractive index.

Resolution: Smallest distance by which two objects can be separated and still be resolved as separate objects.

Segmentation: An image processing technique where a specific characteristic (e.g. grey level) is used to isolate all objects in an image with that characteristic.

Short-pass filter: Optical filter that passes light below, but not above, a certain wavelength.

Signal-to-noise (S/N): Ratio of intensity of signal to that of the background.

Snell's law: Light will bend toward the normal as it passes from a lower to higher refractive index material.

Spherical aberration: Inaccurate focusing of light due to curved surface of lens.

Stokes' shift: Difference between the maximal absorption and emission wavelength of a fluorophore.

Tube length: Distance between back focal plane of objective and primary image plane.

Ultraviolet: Wavelengths of electromagnetic spectrum from 100 to 400 nm.

Wide-field microscope: Microscope in which the full field of view is illuminated.

Working distance: Distance between top of sample and the front surface of the objective, or the bottom of the sample and the top surface of the condenser.

Index

CPSIA information can be obtained at www.ICGtesting.com
Printed in the USA
LVOW05s0826121214

418518LV00003B/11/A

9 781872 748849